Library of Congress Cataloging-in-Publication Data

Integrated environmental management / John Cairns, Jr., Todd V. Crawford, editors.
    p.   cm.
    Includes bibliographical references and index.
    ISBN 0-87371-279-X
    1. Environmental policy—Georgia and South Carolina—Savannah River Region. 2. Environmental impact analysis—Georgia and South Carolina—Savannah River Region. 3. Environmental policy. 4. Environmental impact analysis. I. Cairns, John, 1923- . II. Crawford, Todd V.
HC107.G42S285   1991
363.7'06--dc20   90-20599

COPYRIGHT © 1991 by LEWIS PUBLISHERS, INC.
ALL RIGHTS RESERVED

Neither this book nor any part may be reproduced or transmitted in any form or by any means, electronic or mechanical, including photocopying, microfilming, and recording, or by any information storage and retrieval system, without permission in writing from the publisher.

LEWIS PUBLISHERS, INC.
121 South Main Street, P.O. Drawer 519, Chelsea, Michigan 48118

PRINTED IN THE UNITED STATES OF AMERICA

Integrated Environmental Management

John Cairns, Jr.
Todd V. Crawford
Editors

 LEWIS PUBLISHERS

Preface

Lovelock's Gaia hypothesis, as originally proposed, has sparked much debate as to whether the Planet Earth could or should be treated literally as a living organism. To this industrial manager with an interest in and responsibility for environmental matters, however, the Gaia notion has utility if viewed, metaphorically, as a systems approach for thinking about the interrelatedness of a complicated network containing a myriad of interactive feed-forward and feedback loops with vastly different time frames whose near- and long-term implications are difficult to comprehend separately. A pragmatic view of how to tackle environmental questions is given in *Integrated Environmental Management*, which documents the need, and argues convincingly for, the experimental and economic advantages of using an integrated systems management methodology. The editors are to be commended for putting together a timely handbook that will demonstrate to present and future generations that an integrated ecosystems approach is the right way to manage environmental complexity.

Many of the contributors have had hands-on problem solving experience at the Savannah River Site, especially Ruth Patrick and John Cairns, who have devoted so much of their professional lives to ensuring the health and viability of the Savannah River. It is gratifying that this book captures their dedication.

Herbert S. Eleuterio
Director of Special Studies
Petrochemicals Department
E. I. du Pont de Nemours and Company
Wilmington, Delaware

Foreword

Nuclear weapons and the nuclear industry, nurtured by war, arrived on the scene shortly before the word *environment* became common on the social and political agenda. Both the historical development of the nuclear industry and the unique quality of radioactive materials have contributed to the current environmental situation from which comes the call for integrating environmental management.

A wartime invention, controlled fission to produce energy, was a remarkable technological development after World War II for which there was no real demand. Stimulation of demand became the responsibility of the Atomic Energy Commission (AEC), encouraged by the Joint Committee on Atomic Energy of the United States Congress. Few private institutions or public agencies knew much about the technology of what appeared to be a panacea in the energy field or about the potential hazards to humans and ecology. Some were afraid of a mysterious invisible killer that would last for thousands of years, while some in the public sector simply assumed, or hoped, that a new miraculously cheap source of energy could, without great difficulty, be put to good use. In fact, because the subject was new and complex, most of the debate about the technology was confined to those directly involved within the federal government and associated industry. Moreover, the AEC was both salesman and regulator, a dual role it was not eager to relinquish. Words such as *ecology, risk,* and *biodiversity,* at home in this volume, were known to some at the Savannah River Plant, but were obscure among the public-at-large or within what became the scientific and technical communities concerned with the environment. The field of atomic energy was dominated by brilliant physicists and chemists initially satisfied that help from outside was unnecessary and likely to be misguided.

Experts in the field of sanitary engineering and radiology began the effort through their professional networks to involve the health professions and agencies at the state, local, and federal levels beyond the AEC. While the focus of these experts was on scientific and technical issues, their concern for the health and environment, and the discussion they generated, exposed the issues to a much wider public.[1] It is in this broad public arena with its many actors that most current environmental debates take place. With hindsight, the attempt to build an atomic energy industry within a relatively closed community of participants can be seen as unwise and partly responsible for the mistrust with which it and many new technologies are viewed. This appears to be true despite the fact that the industry has had a comparatively safe history (punctuated by significant failures) when compared with the history of development of a number of other chemical industries. Public participation, disagreement among experts, calls for holistic approaches, intergenerational risk evaluation (the long time scale), and questioning of the "technological fix" all owe much to the arrival of nuclear energy on the social scene.

Abel Wolman, as an expert (a participant in the battle over the safe use of atomic energy) as well as an optimist about the prospects for such use, looked forward to participating in this conference. He would have enjoyed asking the ecologists what exactly they wanted done, while pushing the apostles of the technological future to listen to the ecologists before confidently leaping blindfolded into the future. The dedication to Abel Wolman captures that spirit and the product of the conference testifies to what he missed.

M. Gordon Wolman
The Johns Hopkins University

1. Balogh, B. *Trouble in Paradise: Institutional Expertise in the Development of Nuclear Power, 1945-1975*. UMI Dissertation Information Service (1987).

Acknowledgments

In the preparation and conduct of the conference and the preparation of the book, we had significant assistance from our professional colleagues and from a number of people at our home institutions. We particularly want to recognize the assistance we received from the Savannah River Laboratory personnel. The conference could not have been conducted as smoothly as it was without the able secretarial assistance of Earlis Collins and Andre Gray before, during, and after the actual conference. Conrad Hutto, Nina Baxter, Peggy Glover, and Judy King from the Savannah River Laboratory provided a great deal of input in the planning of the conference and the arrangements that made the conference run smoothly — everything from food to projection equipment. We are indebted to the University of South Carolina at Aiken for allowing us the use of their new Science Building for the conduct of the conference and the Etheredge Center for the Plenary Session. And lastly, material for this book could not have been put together without the able editorial assistance of Darla Donald, Editorial Assistant in the University Center for Environmental and Hazardous Materials Studies at Virginia Polytechnic Institute and State University.

Dedication

The late Abel Wolman was to have been one of the participants at this conference. I am writing this dedication because I had the good fortune of personally knowing Abel for decades, whereas Todd (Crawford) knew him only through their correspondence. Since Abel's untimely death prevented his attendance and sharing his ideas with us, it seems appropriate to acknowledge our respect and affection by dedicating this volume to him.

It is uncommon to use the word "untimely" in connection with the death of a 96-year-old colleague. However, in the academic world, chronological age is less important than intellectual vigor, and in this respect, Abel was still in his prime!

He was a gentle and courteous man. For nearly 20 years, I regularly sent him reprints of each article that I published and invariably received a thank you note shortly thereafter. Just a few months before his death, he mentioned in one of these notes that he was running out of new ways to say thank you. No one else would have had such a thought. For the published material, Abel offered only praise and comments on the good points. However, when I sent him a manuscript to review, I always expected and got a rigorous analysis of the positions I had taken. No deficiency was too small to escape his attention. With all this, the criticisms were always constructive and the manuscript better as a consequence.

Abel was a wonderful companion. Sharing a meal with him was a memorable experience that passed all too quickly. He was always so interested in the activities of others that one had to make a special effort to encourage him to talk about the rich tapestry of events that comprised his life. In an earlier age, he would have been a master storyteller. Even after he passed 70, his keynote addresses focused on the future and the present, mentioning the past only as it related to the present and future. When audiences would have been well satisfied with merely an opportunity to see one of the towering figures in the field, he challenged them. When professionals were in danger of being complacent about accomplishments, he gently but firmly reminded them of deficiencies that cried for attention.

It is fair to say that few people on this planet, past or present, have saved as many lives as Abel Wolman has by developing the disinfection of water with chlorine. Hundreds of millions of people are healthier, even alive, today because of his insights. Moreover, he was as conscious of the new threats to human health and well-being in drinking water as he was of those he helped diminish. Furthermore, he offered many useful insights into solving today's problems as vigorously as he did those of the past.

This is a volume on integrated environmental management, and Abel Wolman is a shining example of the professional activities that exemplify this holistic view of problem solving. He was at home with professionals in a variety of disciplines and regularly read a wide variety of disciplinary journals. Most important, his discussions were never restricted by disciplinary boundaries, as is often the case when interacting with those who make a fetish of their area of specialization.

After Abel Wolman's death, there were many tributes to him in the professional journals and the popular press. Those of us privileged to have known him personally felt a deep sense of personal loss. However, we should celebrate Abel Wolman's life rather than dwell on our personal loss! We can do so by addressing the problems still to be resolved with the same zest and enthusiasm that Abel Wolman showed throughout his career. He undoubtedly endured the blood, sweat, and tears that all professionals encounter, but it is his triumphs we remember. If he experienced defeat, it was forgotten. His unfailing curiosity about the world, his willingness to abandon the status quo for improved methodology and insights, his gentle, unfailing courtesy, and, most important, his constructive interaction with other disciplines and holistic view of world problems are the characteristics that, if utilized by others in the field, would foster rapid resolution of some of the issues identified in this volume.

John Cairns, Jr.

Abel Wolman, 1892—1989
Engineer, Scholar, Educator,
Consultant, Public Servant

American Water Works Association Honors
    AWWA President, 1942
    AWWA Honorary Member
    AWWA Life Member
    Journal AWWA Editor-in-Chief, 1921—1937
    AWWA Medal of Outstanding Service, 1937
    AWWA Distinguished Public Service Award, 1952
    George Warren Fuller Award, 1952
    Water Utility Hall of Fame, 1986

Other Significant Honors and Awards
    Professor Emeritus of Sanitary Engineering, The Johns Hopkins University
    American Public Health Association, Albert Lasker Special Award, 1960
    Research Society of America, Proctor Prize, 1967
    Franklin Institute of America, Lewis L. Dollinger Award, 1968
    The Johns Hopkins University, Milton S. Eisenhower Medal for Distinguished Service, 1973
    National Medal of Science, 1975
    Tyler Ecology Award, 1976
    Rene Dubos Center for Human Environments, Environmental Regeneration Award, 1985
    World Health Organization, Health for All by 2000 Award, 1988
    National Academy of Sciences, 1988

Dr. John Cairns, Jr. is University Distinguished Professor in the Department of Biology and Director of the University Center for Environmental and Hazardous Materials Studies at Virginia Polytechnic Institute and State University, Blacksburg. He received his Ph.D. and his M.S. from the University of Pennsylvania and an A.B. from Swarthmore College, and completed a postdoctoral course in isotope methodology at Hahnemann Medical College, Philadelphia. He was Curator of Limnology at the Academy of Natural Sciences of Philadelphia for 18 years, and has taught at various universities and field stations.

Among his awards are the Presidential Commendation in 1971; the Charles B. Dudley Award in 1978 for excellence in publications from the American Society for Testing and Materials; the Founder's Award of the Society for Environmental Toxicology and Chemistry in 1981; the Icko Iben Award from the American Water Resources Association in 1984; the B. Y. Morrison Medal in 1984; Fellow, American Academy of Arts and Sciences, 1988; the United Nations Environmental Programme Medal in 1988; and the American Fisheries Society Award of Excellence in 1989.

A member of many professional societies, he is a member of the Science Advisory Board of the International Joint Commission. Dr. Cairns has been consultant and researcher for the government and private industries, and has served on numerous scientific committees. His most recent publications are *Rehabilitating Damaged Ecosystems* (CRC Press, 1988) and *Functional Testing for Hazard Evaluation* (ASTM, 1989).

Dr. Todd V. Crawford graduated from California Polytechnic State College in 1953 with a B.S. degree in Agricultural Engineering. He immediately entered the U.S. Air Force in a special program for meteorologists. Under Air Force sponsorship, he attended the University of California at Los Angeles and by 1954 received the equivalent of a Bachelor's Degree in Meteorology. Following 3 years of duty as a weather forecaster in the U.S. Air Force, he returned to UCLA and obtained a Master's Degree in Meteorology in 1958.

From 1958 until 1965, he was on the faculty of the University of California, Davis, where he taught courses in micrometeorology and conducted research in agricultural meteorology and micrometeorology. During part of this time, he also worked on a Ph.D. in Meteorology from UCLA which was completed in 1965.

In 1965, he joined the Lawrence Livermore National Laboratory at Livermore, CA. Here, he developed computer codes for the atmospheric diffusion, deposition, and transport through the environment of radionuclides associated with potential peaceful applications to nuclear explosives. These computer codes were extensively tested against experiments in the late 1960s. In the early 1970s, he was a member of the U.S. delegation discussing peaceful use of nuclear explosives with the U.S.S.R. He was also simultaneously deeply involved in air pollution studies of the Livermore Valley and the San Francisco Bay area.

In 1972, he joined the Savannah River Laboratory which was operated by the du Pont Company for the U.S. government in Aiken, SC. After 1 year as a Senior Research Staff member developing plans for an environmental research program, he became the leader of that program. During the remainder of the 1970s and on into the 1980s, he progressed through successive management positions within the environmental disciplines. In 1986, he was elected a Departmental Fellow of the du Pont Company, one of a very few staff members at this level. In April 1989, the Westinghouse Electric Company took over the operating contract for the Savannah River Site, and Dr. Crawford's role continues to be that of a very Senior Staff person, providing technical leadership to all of the programs of the Laboratory and all of the environmental programs of the Savannah River Site.

Contents

1. Introduction, Todd V. Crawford and John Cairns, Jr. 1

2. The Need for Integrated Environmental Systems Management, John Cairns, Jr. 5

3. Managing Environmental Risks: Ethical Plumb Lines, Margaret N. Maxey 21

4. The Systems Approach to Environmental Assessment, Robert V. O'Neill 39

5. Decision Analysis as a Tool in Integrated Environmental Management, Randall A. Kramer 53

6. Applied Ecology, Its Practice and Philosophy, L. B. Slobodkin and D. E. Dykhuizen 63

7. Clean-Up of Contaminated Lands: How Clean is Clean Enough?, Jerry J. Cohen 71

8. The Savannah River Site as a National Environmental Park, Eugene P. Odum 79

9. A Strategy for the Long-Term Management of the Savannah River Site Lands, F. Ward Whicker 87

10. The Role of the Endangered Species Act in the Conservation of Biological Diversity: An Assessment, Larry D. Harris and Peter C. Frederick 99

11. Endangered Species Protection — The Wood Stork Example, William D. McCort and Malcolm C. Coulter 119

12. The Savannah River — Past, Present, and Future, Ruth Patrick 137

13. Impacts of Management Decisions on Environmental Issues of the Savannah River, Paul B. Zielinski 151

14. The Savannah River System as a Microcosm of World Problems: Instructions to Conference Participants, Kenneth L. Dickson 157

15. Management of the Savannah River, Paul Zielinski, Bernd Kahn,
    B. Badr, P. Cumbie, J. Dozier, J. Gordon, R. Lanier, M. Parrott,
    R. Patrick, D. Sheer, R. Vannote, and N. Weatherup 161

16. Endangered Species Protection — The Wood Stork,
    William D. McCort, Nora A. Murdock, I. L. Brisbane,
    F. B. Christenson, M. C. Coulter, J. R. Jansen, R. A. Kramer,
    S. Loeb, H. E. Mackey, T. M. Murphy, and P. Stengle 175

17. Long-Term Management of the Savannah River Site Lands,
    F. Ward Whicker, Jerry J. Cohen, S. Bloomfield, K. L. Dickson,
    E. C. Goodson, J. G. Irwin, M. N. Maxey, R. V. O'Neill, J. F. Proctor,
    L. E. Rodgers, and S. R. Wright 179

18. Future Needs, John Cairns, Jr. 183

19. Summary of Perspectives on Integrated Environmental Management,
    Laurence R. Jahn 193

Appendices 199

Index 211

CHAPTER 1
Introduction

Todd V. Crawford and John Cairns, Jr.

During the 1960s and the early 1970s, the United States population became much more aware of environmental issues. Universities increased their focus on environmental sciences, and legislatures began passing a significant number of environmental laws. These early and subsequent laws resulted in a significant improvement of the environment in the United States. Over the last decade (the mid-1970s through the 1980s), promulgation of environmental laws and regulations has increased significantly, as is illustrated in Figure 1. Concurrently, the focus of legislators, politicians, regulators, and academicians has become much more narrowly defined. In the environmental regulatory field, for instance, laws are becoming much more prescriptive, and in academia, the higher one pursues an education the more narrowly focused it becomes.

There is currently little latitude for judgment in environmental matters, yet the flexibility to utilize judgment is required when the health of the total environment is the issue of prime concern. Instead, the United States culture has moved toward more detail in the environmental regulatory field. This has been encouraged by an adversary-based judicial system. In such a system, there is little opportunity for optimally resolving complex issues. A number of examples can be identified where environmental management decisions have been made to conform to a narrow aspect of a particular law that has been detrimental to the environment when considered holistically.

The genesis of this conference and resulting book was a dinner between us in December 1987 in Augusta, GA. We were waiting for Dr. Ruth Patrick to arrive at the Augusta Airport for a regularly scheduled meeting of the Environmental

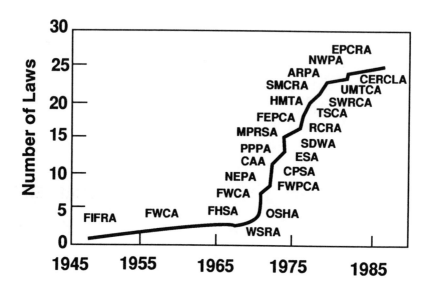

FIGURE 1. Legislation related to environmental protection. Some of the major federal laws applicable at SRS are CAA = Clean Air Act; CERCLA = Comprehensive Environmental Response, Compensation and Liability Act; CWA = Clean Water Act; ESA = Endangered Species Act; FIFRA = Federal Insecticide, Fungacide, and Rodenticide Act; HMTA = Hazardous Materials Transportation Act; NEPA = National Environmental Policy Act; NHPA = National Historic Preservation Act; OSHA = Occupational Safety and Health Act; RCRA = Resource Conservation and Recovery Act; SARA = Superfund Amendments and Reauthorization Act; SDWA = Safe Drinking Water Act; TSCA = Toxic Substances Control Act.

Advisory Committee (of which we are all members) for the operating contractor at the Savannah River Site (SRS). This was Du Pont in 1987 and is currently Westinghouse. At this dinner, the two of us discussed the need to have environmental management decision makers think broadly on such issues. In order to focus attention on this subject, we conceived the idea of having a conference on Integrated Environmental Management. At the conference we would form working groups to tackle multidisciplined environmental management problems with the aim of obtaining some consensus approaches to solving such problems. Associated with the conference, we would have a series of keynote speakers and we would publish a book.

PURPOSE

The purpose of the conference was to develop approaches for managing complex multidisciplined environmental situations. This development would be done in such a way that the interactions between the different environmental media and uses would be considered, that all participants would contribute, that input from the keynote talks would be utilized, and the working group discussions would focus on the expected products. The purpose of this book is to have a positive impact on the way complex environmental management issues are resolved in the United States. This could be accomplished through a consideration of the environmental educational system, through changes in the regulatory posture and laws of the United States, through the way environmental issues are resolved in the court system, etc.

Three topics were selected for the working groups to be of local interest to the Savannah River-South Carolina-Georgia area:

1. Management of the Savannah River
2. Endangered Species Protection - The Wood Stork Example
3. Long-Term Management of the Savannah River Site Lands

Each of these working group topics was initiated with a principal question. Illustrative questions were also furnished to all the participants prior to the conference to focus attention on the issues to be addressed (questions are stated in Appendix I of this volume). Even though a specific focus was selected for the discussions, the lessons learned have applicability to other river basins, other endangered species, and other large federal land holdings.

CONDUCT

The conference on Integrated Environmental Management was conducted at the University of South Carolina campus in Aiken, SC, September 26 to 28, 1989. It was hosted by the Savannah River Laboratory with support through the Department of Energy, Savannah River Operations Office. Speakers were selected for their ability to contribute background information and motivational thoughts to the main purpose of the conference. Participants were selected for their interest and ability to represent a wide variety of views in the three working groups. Selections and invitations were all issued from the convenors, although input was received from a number of colleagues, and particularly the chairmen of the three working groups. The intent was to have all manuscripts in hand by August 1989 so they could be distributed to the conference participants early in September. With only one or two exceptions, this was accomplished, and the manuscripts were distributed ahead of the conference to all of the participants.

Chapters 2 to 14 are talks given at the meeting, with a couple of exceptions. The paper by L. B. Slobodkin and D. E. Dykhuizen (Chapter 6) was not presented personally at the meeting, but is offered here and was offered to the participants ahead of the meeting. At the last minute, E. P. Odum was unable to attend the conference, but he has prepared Chapter 8. Ian McHarg gave a presentation, "Land Use Planning for Multiple Uses," at the conference, but did not provide a manuscript for the book.

The three papers by Paul Zielinski (Chapter 13), William McCort and Malcolm Coulter (Chapter 11), and Ward Whicker (Chapter 9) posed the environmental management issues, along with the presentation of background information for the working group study cases. These were also distributed ahead of the meeting. The paper by K. L. Dickson (Chapter 14) was the last formal prepared talk, and it was given as a charge to the working group participants.

Participants were divided into three working groups, each with a chairman and a rapporteur.

At the conclusion of the conference, a plenary session was held in the Etheredge Center of the University of South Carolina, Aiken Campus, which was open to a much larger number of local interested people. At the plenary session, John Cairns summarized the keynote talks (Chapter 18), and the three working groups summarized their findings (Chapters 15, 16, and 17). The overall conference rapporteur was Lawrence Jahn. At the plenary session he pulled together the lessons learned from the entire conference (Chapter 19). The material prepared at the conference was submitted in final form in early November 1989, whereas the material that was prepared ahead of the conference was largely in final form at the conference. The complete list of participants, along with addresses and phone numbers, is listed as Appendix II of this volume.

CHAPTER 2

The Need for Integrated Environmental Systems Management

John Cairns, Jr.

INTRODUCTION

Integrated environmental management (IEM) may be defined as proactive or preventive measures that maintain the environment in good condition for a variety of long-range sustainable uses. Alternatively, IEM may be regarded as coordinated control, direction, or influence of all human activities in a defined environmental system to achieve and balance the broadest possible range of short- and long-term objectives (I am indebted to Daniel Scherer for this definition). Sometimes a course of action is clearer by stating what it is not. Environmental management is NOT fragmented decision making so that only one use is considered at a time. Short-range goals that benefit a single group are NOT preferred.

The benefits of IEM are so evident, after even the most superficial examination, that one wonders why implementation appears so difficult. The more obvious benefits are:

1. Long-term protection of the resource
2. Enhanced potential for nondeleterious multiple use
3. Reduced expenditure of energy and money on conflicts over competing uses and the possibility of redirecting these energies and funds to environmental management
4. More rapid and effective rehabilitation of damaged ecosystems to a more usable condition (more ecosystem services provided)
5. Cost effectiveness

Resource managers have long recognized that the institutional problems associated with managing natural resources are invariably more aggravating and intractable than the scientific and technical problems. For a list of such questions and problems, see the Addendum to this chapter. Typically, lack of methodology is not what impedes more effective use of natural systems (although methodology could certainly be improved), but rather many institutions (each charged with a fragment of resource use or management) fail to integrate system management responsibilities. This situation was created as part of national policy when specialization was the dominant theme in science and engineering. The field of medicine has been redirected recently toward a holistic approach, even though specialists still play a critical role. Environmental management must do the same.

The tremendous cost to society of fragmented environmental management is not fully appreciated. Resource management strategies developed by specialists often cannot be implemented effectively because they are in conflict with other strategies developed by specialists with entirely different perspectives. Academic blocks are also present in examining environmental management alternatives.[1]

ABANDONING THE FRONTIER ETHIC

Homo sapiens has been at war with nature throughout its history. In prehistoric times, humans fought being eaten by large predators. In recent times, man has fought other mammals (e.g., rats), insects, and the like for food, to prevent damage to buildings and livestock, and for a variety of other comparable purposes. In other words, certain "pests" were competing for food that man valued and were damaging livestock, buildings, etc. Although insect and other animal and plant pests are far from controlled, technology and machinery have advanced sufficiently to damage nature on a larger scale and with a more devastating effect than ever before in the history of the human species. While winning every battle with natural species is not possible, man is "winning" the war with nature through large-scale destruction of habitat, etc. This attitude of warring with rather than wooing nature must be discarded.

THE PRESENT NEED FOR INTEGRATED ENVIRONMENTAL MANAGEMENT

At the time many of us still alive were born, vast game herds roamed Africa, sizable forests untouched by chainsaws or even axes existed in parts of North America, the tropical rain forests in South America and elsewhere were relatively undisturbed, and even China's teeming population was less than half its present size. In the era before that, human populations were thinly distributed over the earth; centers of population did exist, but large stretches of relatively pristine natural systems also co-existed. Today, pressure on natural resources is intense almost everywhere. Natural resources are no longer capable of satisfying every perceived need.

I am preparing this at the Rocky Mountain Biological Laboratory on the western slopes of the Rocky Mountains in Colorado. The demands on the surface water supply here are a constant source of contention. Colorado's large centers of human population are on the eastern slopes and large quantities of water are in the relatively thinly populated western slope areas, flowing "unused" to other states. The developers of Aurora, CO, want the water from the Gunnison drainage to be used for further development of their city. Ranchers on the western slopes want this water for irrigation; other uses (often competing with the ranchers) are fishing, rafting, and other recreational uses. Not only is the quantity allocation important, but some uses, such as extensive irrigation, change the water quality so that even the portion that returns to the natural system is altered in ways often deleterious to indigenous biota.

Going further afield, rather solid evidence shows the buildup of atmospheric carbon dioxide; less solid but persuasive evidence concludes that this and other gases are contributing to global warming. Some of this carbon could be stored for relatively long periods of time in the Brazilian rain forests if they were not being destroyed at an unprecedented rate. Destruction of these forests through burning is contributing still more carbon dioxide to the atmosphere. The fisheries of the world's oceans are being overexploited, and, in many instances, the breeding stock is seriously threatened. All these furnish a multiplicity of warnings that the resource base upon which life support depends is being threatened and will not be suitable for sustained use.

IEM should make sustained use possible. If the frontier attitude is not abandoned, the earth may soon be far less habitable.[2] Persuasive evidence indicates that with equal distribution of food and no grain feeding of animals, the 1985 global harvest would have been sufficient to feed an essentially vegetarian diet to 6 billion people.[3] It is only possible to support the earth's present population of roughly 5.2 billion people because society collectively is doing something no prudent family would consider: squandering its inheritance.[2] Resource capital such as fossil fuels, topsoil, major forests, and groundwater are being used for daily living. Perhaps most important, humans are responsible for destroying fellow species at an unprecedented rate.[4,5] This is not environmental management — it is environmental exploitation! Man may have as little as 10 years to change his ways drastically.[6] To move from global environmental exploitation to environmental management to integrated environmental management in 10 years is a formidable social and political challenge. Yet if man does not, society as he knows it will probably not survive.

The stability of the social/political system is extremely fragile, and a major crop failure even in part of the world (e.g., Egypt, India, or mainland China) might well destroy the present uneasy equilibrium. Even if no crops failed due directly to global warming (highly unlikely), rising sea levels would indirectly reduce food supplies through flooding, salt contamination of coastal aquifers used for irrigation, and destruction of coastal wetlands that serve as fishery "nursery" grounds, etc.

Of course, it may not matter if half the world's species are lost, and global warming may not result from anthropogenic changes in the atmosphere or the situation may be benign if it does occur. Similarly, changes in the ozone layer may improve our complexions. Then again, we might find that they are a threat after the time has passed for effective remedial action. Why not stop these trends until the information base is larger? We can always revert to our present ways if the evidence shows these changes are good for the planet. (We may even eliminate the United States budget deficit as presidents have promised for years.) The bottom line is that present natural resource practices cannot persist much longer, and plans should be made for practices that are compatible with sustained, nondegrading resource use.

Why are we playing this dangerous game when compelling evidence shows that we cannot win much longer? Ornstein and Ehrlich[7] conclude that the problem is basically evolutionary. Except for the most recent portion of human history, no reasons have surfaced for evolving, either biologically or culturally, a capacity to detect and respond to gradual trends in the environment. In fact, even if our remote ancestors had detected gradual climate change, nothing could have been done to alter it. This is no longer true. Anthropogenic changes can often be detected in the environment, and we should be capable of responding to those that may affect us adversely. This might be categorized as *defensive environmental management*.

DEVELOPING A MANAGEMENT PLAN

An effective management plan cannot be developed without an extensive and ongoing knowledge of the condition of the system being managed, including its important components. Of course, death of salmon and sea otters in Prince William Sound, Alaska, indicates that the system is not being managed properly. Similarly, the recent appearance of large quantities of hypodermic needles and other hospital wastes on bathing beaches is an unmistakable sign of poor management practices. However, management consists of doing more than correcting gross and highly visible deficiencies. Good management consists of receiving an early warning of malfunction and correcting deficiencies before they become as serious as those just described.

Continual surveillance of the natural resource systems that collectively comprise the earth's life support system will be a new cost that many members of society will be reluctant to pay. In the past, natural resources have been regarded by most as "free goods," defined by economists as those in unlimited and inexhaustible supply for which the demand cannot exceed availability. However, ample evidence has already shown that natural resources are no longer in an unlimited supply and that the demands upon them are much greater than they are capable of enduring for long-term sustainable use. In short, natural resources are no longer free goods. The only question is how to manage them to maximize their benefits to society for long-term sustainable use and, equally important, to determine who pays the management costs. Fragments of this larger responsibility have already been assigned to various government and state regulatory agencies, such as the U.S. Army Corps of

Engineers, the U.S. Environmental Protection Agency, the Department of the Interior, the National Atmospheric and Oceanographic Agency, and such private organizations as The Nature Conservancy. However, fragmented management of a limited component of a larger system is unlikely to work well at best, and will place the various management groups in conflict with each other at worst. Conflicts of this sort rarely place the well being of the resource as the primary goal. In short, present, rather primitive, environmental management responsibilities are severely fragmented, often with little or no cooperation within a single governmental jurisdiction such as the United States and less cooperation internationally even between such good neighbors as the United States and Canada. In international situations, such as managing the whale stocks in the ocean, a number of nations have flagrantly abused cooperative agreements by simply ignoring them or by using the guise of research.

I had once thought that the problem was a lack of perception by key administrators, but presenting the keynote address at the Second Natural Resources Leaders Seminar, Charleston, SC, December 12 to 14, 1985, altered this opinion. I quickly learned at this meeting that the top administrators (the head and four chief assistants) of each of the governmental agencies who were present for the entire workshop clearly recognized the need for IEM and the benefits to be derived from adopting such a policy. The big question was implementation.

One of the important issues not usually frankly and openly discussed fits the category of "turf battles." An agency's power and authority are often determined by size and areas of responsibility. No agency wants to lose either of these. This concern is not limited to administrators, as staff members in agencies are also concerned about job security. Losing any "turf" means some activities of their agency are likely to be taken over by another agency or even shared by another agency. Until these issues are resolved, integrated environmental management will not be embraced.

The way Congress and state legislatures assign funding is also not likely to encourage IEM. In this era of accountability, performance can be more readily assessed if funding is given for a specific mission, especially when there are ways of determining whether the mission has been carried out successfully. In some cases, however, the mission of one agency to control flooding may not be compatible with the mission of another agency that is assigned responsibility for preservation of bottomland hardwood ecosystems that are dependent upon periodic flooding. Such conflicts are also present in the public health area when one agency is espousing reduction in cigarette smoking to improve human health and another agency is encouraging tobacco farming as part of a diverse agricultural system. IEM cannot eliminate these conflicting responsibilities, but it would certainly call attention to them. Once enlightened management becomes aware of these problems, they would at least be addressed more effectively.

USE OF DECISION ANALYSIS

Decision analysis can be used in resolving some of these conflicts[8] and, more important, in identifying the issues that are likely to polarize opinion, often before opinions actually become polarized. Once when I was testifying before a congressional committee, the committee chair accused me of wanting to establish "an environmental czar." That was not my intention, since an environmental czar could presumably use the environment as he/she saw fit. Actually, I do favor making the maintenance of environmental condition the dominating force since that is the only way long-term sustainable use is possible. The organization charged with this responsibility would indeed acquire substantial power, but the power should not be used capriciously or in the interest of the agency, but rather in the maintenance of environmental quality and condition compatible with sustainable use.

PAYING FOR INTEGRATED ENVIRONMENTAL MANAGEMENT

This discussion probably seems bizarre in terms of present practices, but alternatives to the present practices (which are inadequate) must be developed and considered. If the environment is no longer considered a "free good," then allocating its use (i.e., nondegrading assimilative capacity) should have some economic basis. The money charged for its use would be used to ensure that the biological integrity of the ecosystems being used was not impaired. I earlier[9] defined biological integrity as the maintenance of structure and function characteristic of a particular locale. All environmental use must be based on the assumption that the environment has a certain assimilative capacity[10,11] for societal wastes. That is, the environment is capable of receiving certain wastes at certain levels for particular periods of time without suffering deleterious effects or having biological integrity impaired. This assumption has been vigorously challenged by Campbell,[12] who disagreed with this idea. The Campbell article was rebutted by Westman[13] and Cairns.[14] A more detailed examination of this hotly debated assumption of assimilative capacity might be interesting reading for those in the field. However, the debate was primarily academic, and the matter can be settled quickly in practical terms. *If the assumption is made that the environment has no assimilative capacity (that is, any use of it will cause deleterious change), then human society as it now exists cannot be sustained!* Whatever the academic merits of the debate, it is quite clear that politicians and the general public do not wish to bring the present social system to a dead halt and reduce the population density to prehistory levels. The question then becomes not an academic one but rather how much societal waste will impair ecological integrity. Persuasive evidence indicates that nondegrading use is possible, although the evidence is definitely not conclusive. Even more persuasive evidence shows that the present assimilative capacity of natural systems is being exceeded by a wide margin and that current practices are not suitable for sustainable environmental use.

In riverine ecosystems, river basin management groups, such institutions as ORSANCO, Thames Water Authority, Delaware River Basin Management Commission, and Potomac River Commission, are good examples of a partial overall management group. However, none of them have the authority to raise funds or to limit the use of the riverine ecosystem, both of which I am espousing here.

EXCHANGE OF INFORMATION

If groups with diverse professional backgrounds, diverse objectives, and frequently radically different missions are to work together effectively, these differences must be understood and the latitude in achieving operational conditions must be explicitly stated. For example, one organization using the Savannah River might want an adequate supply of cooling water, another may be assigned the responsibility of reducing or eliminating floods, yet another must maintain channels at a certain depth, others may wish to use the river for assimilation of municipal wastes or dilution of industrial wastes, and still others may be interested in recreational activities. Even this limited list of multiple uses of the Savannah River shows that optimizing one use is often in clear and direct conflict with optimizing one or more of the other uses. Each organization naturally wishes to optimize its particular beneficial use. The organization might, either unwittingly or deliberately, attempt to do so and, as a consequence, impair beneficial uses of others. In most cases, a range of use is acceptable, even though not optimal. This permits greater flexibility in arranging multiple environmental use than when each user attempts to achieve optimal utilization of the resource for a single purpose.

IEM will almost certainly mean reduced use if the resource is overutilized (i.e., cumulative impact may impair biological integrity even though each use is individually acceptable). However, reduced use does not necessarily mean proportionally reduced benefits to the user. For example, an industry discharging raw waste into a river is using it for all the steps in the waste treatment process. Primary and secondary treatments of raw waste are the cheapest, and the tertiary treatment (final polishing) is the most expensive. However, the final polishing puts the least strain on the assimilative capacity of the receiving system. Such considerations are very important in IEM. Industries with such small profit margins that they cannot provide even basic waste treatment are probably not going to survive in the long run. Facing this problem immediately may be best for the industry and for the area that benefits from the taxes it provides. Without doubt, such restrictions may either make the proposed use economically unfeasible or the persons who must limit this use may feel that their "rights" to use the river are being unduly restricted for reasons not entirely clear to them.

Very often, the objectives and operational prerequisites of others who are using a resource are not well understood by those attempting to make another, often conflicting, use of the resource. Alternatively, restrictions may be well understood, but Congress, state legislators, or corporate executives have charged the regulators with a particular responsibility and related the funding of the agency or organization to the degree to which this particular objective is achieved. This situation will almost certainly lead to a number of conflicts, which is exactly what is happening all too commonly with resource use today. Often, in attempting to achieve these narrow objectives, the collective impact on the river or other ecosystem may be so severe that it will not be suitable for long-range sustainable multiple use.

The first two steps in IEM are to be certain that each specific objective in resource use is understood not only by those charged with this use but by others as well (Table 1). This includes other organizations attempting to use the resource, legislators, and

Table 1. Protocol for Integrated Environmental Management

Step 1:	Identification of: a. The system limits b. The geographic boundaries, etc. c. History, present condition, and alternative futures of the resource
Step 2:	Identification of: a. All organizations intending to use the resource b. Uses, including episodic, by the general public c. Potential impacts of proposed uses outside the management area
Step 3:	Inform all interested parties, including the general public, of the entire spectrum of organizations wishing to use the resource
Step 4:	Require each organization proposing to use the resource to indicate how the proposed use would affect the resource
Step 5:	Send this information to all resource users - identify conflicting or damaging uses
Step 6:	Use decision analysis or comparable techniques to resolve conflict situations, including activities not compatible with long-range sustainable use
Step 7:	Establish quality control conditions to ensure that the resource is not damaged by proposed uses
Step 8:	Implement monitoring program to ensure that predetermined quality control conditions are being met

the general public. Legislators may often have line items in different places in the budget that, in fact, pit one organization against another, even though both receive funding from the same source.

Of course, each resource user (used in the broad sense of anyone benefiting from the resource) must make an individual determination since each is driven by different sets of goals, objectives, and constraints. Nevertheless, while the determination of objectives must often necessarily be individual, objectives are not independent of each other since all involve a resource for which others are making similar judgments. In my own field of ecotoxicology, antagonistic effects surface when one component works against the other and reduces its effectiveness. In addition, synergistic interactions may be facilitating or enhancing. These negative and positive interactions simply cannot be ignored by a rational decision maker.

Inadequacies in IEM are often simply a failure of "institutional memory." Some events, such as hundred-year floods and other widely spaced episodic disturbances, occur so irregularly and at such widely spaced intervals that persons actually experiencing these while in a particular organization leave for another position, retire, or are promoted to some other responsibility in the organization well before the next devastating event occurs. As a consequence, new personnel think that the present situation is the rule and forget the exception, though it may be devastating.

In the case of truly devastating floods, private citizens may rebuild on the flood plain even immediately after a flood, and local authorities permit them to do so. Even when authorities attempt to restrict building on a flood plain, a few years of benign conditions soon give rise to a false sense of security or authorities are replaced by others who have not personally experienced the devastation of a major flood. As a consequence, building on the flood plain becomes accelerated, and the next flood brings even more damage and a greater outcry for protection. Building flood protection dams causes a number of difficulties, including a marked readjustment of biota,[15] and may severely restrict other beneficial uses. In some cases, flood-dependent ecosystems, such as bottomland hardwoods, are wiped out; organisms such as the wood stork that feed in flood plain pools are threatened or endangered; toxic materials may accumulate in bottom sediments because the system is not periodically cleansed by vigorous natural floods; and the other dynamics of a river, such as cutting new channels, etc., may be eliminated or severely impaired. All of these artificial conditions are the result of single objective environmental management that would be unnecessary if IEM restricted use of the flood plain to activities not strongly adversely affected by flooding. Society would also not subsidize those who tried to get "cheap" land on the flood plain and need continual tax dollars subsidizing their use of it.

BARRIERS TO INTEGRATED ENVIRONMENTAL MANAGEMENT

My attention was first drawn strongly to the barriers to IEM in academe when I lost a valued research assistant because his credentials were considered inadequate for promotion from associate to full professor in a particular discipline, although everyone admitted that they were exemplary in the larger field of environmental science. This incident was much on my mind when I was asked to contribute a chapter to the Thames/Potomac Seminars Series.[1] The word "resource" was then considered interchangable with the word "environment," although some people have pointed out that human talent might also be considered a resource and the word "environment" should be used instead. At a meeting of the Second Natural Resources Leaders Seminar, I presented the justification for abandoning fragmented decision making on environmental problems and espoused an integrated holistic view.[16] At the end of my presentation, I was braced for hostile attack and instead found to my astonishment that everyone agreed with what I had said: that implementation was the main problem. It is evident that talking and thinking about system level integration is intuitively reasonable, but that implementation of such thinking is immediately blocked by a number of barriers. A brief illustrative list of 24 such barriers follows:

1. Institutions of higher education are primarily reductionist, not integrative. This is an age of specialization, and reductionist science and engineering is

responsible for most of the technological advances that have both improved the quality of life and caused it to deteriorate. The scientific method is basically reductionist and, of course, hard science, the most admired of all, is also the most reductionist and quantitative. However, despite the fact that reductionism is necessary to find out how things work, integrative science is necessary to put the new discoveries in a larger context and also to determine what the new inventions and methods will do in addition to what they are designed to do. For example, nuclear power was once regarded much more favorably for the simple reason that reductionist science and engineering told us how it would work, and it did, but not what some of the side effects would be. As these frequently unpleasant discoveries began to surface, the products of reductionist science became much less attractive to the general public. However, the drawbacks could have been seen much earlier, and perhaps a more balanced view toward nuclear power could have been developed. Even now, integrative science shows that the second generation of nuclear power plants may actually do less environmental damage than atmospheric carbon dioxide from fossil fuel power plants that contribute to global warming.

2. IEM takes time. Time means money, and most agencies and institutions do not fund time necessary for developing a sound IEM strategy. Until the practice of funding this essential management activity becomes more common, it is unlikely to occur to the degree that is needed.

3. Turf battles run rampant in organizations. Integrative science should flourish in academe because holistic scholars are generally highly regarded. Nevertheless, at the operational level, departments representing specific disciplines jealously guard their "turf" in types of classes to be taught, degrees awarded, etc. New fields such as biotechnology transcend traditional disciplines, and the relationship among departments that have rarely, if ever, worked together is uneasy despite the fact that individual investigators may work together with few reservations. Credit is difficult to divide; block funding is always contentious when it must go to more than one department, etc.

4. Job insecurity is of paramount importance to all. Inevitably, IEM will reveal that several agencies may be collecting similar data in similar locations at roughly similar times. One of the benefits of IEM is eliminating data redundancy where such redundancy does not help improve the decision making. In some cases, validation of results is necessary; in other cases, there may be more data than are necessary for confirmation or validation that are therefore adding nothing to the usable information base. Awareness of this likelihood makes some employees apprehensive because it could easily be their job that becomes unnecessary. Reassurance that they will have equally interesting assignments or that the data gathering effort can be pooled, so that one agency does not gain ascendancy over another, etc., should mitigate these

fears. These are, nevertheless, real fears that may not always be expressed openly but which must be given attention. The same thing is true, of course, in universities where it is feared that interdisciplinary centers will take students away from the more traditional disciplines, especially when the professional marketplace favors interdisciplinary activities. Therefore, control over the graduating process is often denied interdisciplinary centers or departments on various campuses. It is when the loss of faculty positions where assignments of faculty to specific departments or centers is formula driven (another form of job insecurity) that such problems become exacerbated.

5. Many are unwilling to compromise. Extremists, both among the environmentalists and among the industrialists, take the position that any compromise is totally unacceptable and that the contending group must come around to the "in" group's view before any interaction can be considered justified. The same hostility often occurs among disciplines, especially when hard scientists are mixed with soft scientists. These sorts of "ego trips" are totally unacceptable to IEM where the assumption is that a diversity of viewpoints must be considered and that no one particular point of view should dominate.

6. Short-term profits are too enticing — whaling, tropical rain forests.

7. There is a "what has posterity done for me" attitude.

8. Issues are not simple "good guys/bad guys" decision makers; IEM is hard work.

9. The uncertainty of the outcome is often unacceptably high. Predicting the precise environmental benefits of reducing stack emissions from fossil fuel power plants is a very risky business. This is one of the reasons there is a reluctance to take any action. It is presumed that cutting the deleterious materials in the stack emissions in half, as has been proposed, would be very costly and the biological benefits not entirely clear. The same thing is true of efforts to restore the tropical rain forests. In both situations, the problem is that the science of ecology still does not have the robust predictive models needed to make the outcome of a particular course of action more certain. This makes IEM all the more difficult, but it is nevertheless essential that it be attempted even where the outcome cannot be precisely predicted because the present course of action is clearly not sustainable over a long period of time, and natural ecosystems are badly overloaded over most of the earth's surface. Problems such as enormous loss of topsoil, prospects of global warming, and storage of hazardous waste materials, will very likely have such severe consequences that by the time the information base is adequate for the construction of a robust predictive model, it may be too late to take corrective

action. A 95% confidence level is not necessary for decision making. If, for example, one were offered to select 1 from 20 cups of coffee, one of which contains poison, the decision would probably be not to drink any of them. A decision would be made on the basis of a low probability of deleterious effects. Consequently, a high percentage of confidence was not necessary to make this decision.

10. At the global level, developing countries aspire to material benefits per capita now enjoyed by developed countries.

11. Changes in lifestyle required (recycling, etc.) are strongly resisted by some.

12. Specialists feel more comfortable working with "their own kind" rather than alien (other disciplines) specialists.

13. Environmental despoilers fear that the general public will not have the same value system (e.g., clear-cutters prefer lobbyists to trying to persuade the general public).

14. The present use is considered a "right" not open to discussion or compromise (e.g., prior appropriation water rights of western states, such as Colorado or Native American harvesting of salmon).

15. Society is oriented toward growth rather than maintenance.

16. Change is only acceptable in a crisis, not when the problem is more manageable (e.g., protecting hospital staff vs. the right of the individual to decline an AIDS test).

17. There is a general fear that management authority will be abused.

18. Others fear peer criticism of oversimplification.

19. The belief that all systems are too complex to permit any prescriptive (read: *standard methods and procedures*) legislation or professional endorsement halts some scientists in their progress.

20. People turn off when they face complex issues.

21. Technical information is inadequate.

22. Nonspecialists have difficulty determining which evidence is credible.

THE NEED FOR INTEGRATED ENVIRONMENTAL SYSTEMS

23. The number of professionals skilled in integrated environmental management or some component of IEM is inadequate.

24. The political process is oriented toward polarized issues rather than IEM.

CONCLUDING STATEMENT

There is heartening evidence that the need for integrated environmental management has been recognized.[17] However, a vast gulf can exist between recognizing a need and wide-scale implementation of a policy to meet that need. Some major steps in reducing this gulf are:

1. Endorsing integrated environmental
2. Supporting the move from fragmented to integrated environmental management
3. Focusing on ecosystem attributes rather than management by component (e.g., by species or by single purpose objective)
4. Helping identify the important thresholds for estimating ecosystem assimilative capacity; these are essential to any natural ecosystem (resource) quality control system
5. Avoiding fragmented, incremental decision making, thereby helping to assure that resources yield multiple benefits on a long-term sustained use basis
6. Improving both the quality and quantity of the present resource base by encouraging rehabilitation of damaged ecosystems
7. Recognizing that restoration to the predisturbance condition may not be possible, but an improved ecological condition usually is possible
8. Encouraging educational institutions to facilitate interdisciplinary activities, including IEM
9. Encouraging employers of graduates from academic institutions to emphasize the need for interdisciplinary experiences as students
10. Conducting surveys to gain an understanding of citizen attitudes toward environmental management and support for integrated environmetnal management in particular.

National parks, nature preserves, wildlife refuges, and other protected areas total less than 3% of the Earth's land area.[18] The rest of Planet Earth's life support system is at risk unless *integrated environmental management* is used to maintain ecological integrity for long-term sustained use. IEM will not receive the attention it deserves as long as it remains a fuzzy abstraction in the minds of government officials, industrial executives, state and federal agency personnel, and the average citizen. Unless ecological integrity is protected, the global deterioration now underway will almost certainly destroy society as we know it. The challenge is to develop effective yet practical IEM policies that permit sustainable multiple use of the planet's resources.

ADDENDUM*

The following questions relate to both the Savannah River and to ecosystems in general. In one way or another, they illustrate the need for an ecosystem perspective coupled with an IEM policy. Development of an IEM program for an ecosystem would facilitate having answers to all of these questions available:

1. What are the existing barriers that have prevented an ecosystem perspective on toxic substances and their management?
2. Can information obtained from monitoring an ecosystem to assure that previously established quality control limits are being met also be used as "an early warning system" to protect humans?
3. Is there existing methodology for estimating the multiple exposure risk to target populations within an ecosystem and is there a management protocol for taking corrective action based on this information when necessary?
4. What effects, if any, result from prolonged ingestion of fish from water containing trace levels of toxic chemicals or other hazardous substances?
5. What research methods are available to quantify the different patterns of toxic exposure risk to the environment and human health?
6. How can reactive interest of citizens opposing environmental degradation in their own back yard be converted to proactive efforts to preserve the environment as a whole?
7. What are the benefits and costs (including concealed costs) in ignoring long-term burdens to society for the sake of short-term gains with respect to economic exploitation of natural resources?
8. Are existing institutional frameworks adequate for development and appropriate interpretation of risk from toxic chemicals and other harmful substances for the Savannah River and for the management of biological, physical, and social dimensions of these risks?
9. Are there significant differences in interpretation regarding how risk is communicated by regulatory agencies to the general public and how risk is perceived by citizens?
10. How can scientists do a better job of communicating risk so that the public will support sound decisions leading toward IEM?

---

* These questions were greatly influenced by similar questions raised by the Great Lakes Science Advisory Board of the International Joint Commission, on which I am privileged to serve. Colleagues in that group and others associated with the International Joint Commission have markedly influenced my view of these problems to such a degree that it would be impossible to view problems in other aquatic ecosystems without acknowledging this influence. The modifications in the questions are, of course, my own, and I take full responsibility for them.

11. IEM is probably the only way to prevent the level of environmental degradation that precludes long-term sustainable use. How can the concept of long-term sustainable use effectively be communicated to the general public?
12. What implications for IEM are there in considering people (especially local populations) as part of the natural ecosystem?
13. How can we develop a better citizen understanding of ecological effects in order to encourage support for IEM, responsible individual behavior, and to generate political support for legislative action encouraging IEM?

REFERENCES

1. Cairns, J., Jr. "Academic Blocks to Assessing Environmental Impact of Water Supply Alternatives," in *The Thames Potomac Seminars,* A. M. Blackburn, Ed. (Bethesda, MD: Interstate Commission on the Potomac River Basin, 1979), pp. 77-79.
2. Ehrlich, P. R. "AIBS News: Facing the Habitability Crisis," *BioScience* 39(7):480-482 (1989).
3. Kates, R. W., R. S. Chen, T. E. Downing, J. Kasperson, E. Messer, and S. R. Millman. *The Hunger Report: 1988* (Providence, RI: Alan Shawn Feinstein World Hunger Program, Brown University, 1988).
4. Cairns, J., Jr. "Can the Global Loss of Species be Stopped?" *Spec. Sci. Tech.* 11(3):189-196 (1988).
5. Wilson, E. O. *Biodiversity* (Washington, D.C.: National Academy Press, 1988).
6. Lovejoy, T. E. "Will Unexpectedly the Top Blow Off?" *BioScience* 38(10):722-726 (1989).
7. Ornstein, R., and P. R. Ehrlich. *New World/New Mind: Moving Toward Conscious Evolution* (New York: Doubleday, 1989).
8. Maguire, L. A. "Decision Analysis: An Integrated Approach to Ecosystem Exploitation and Rehabilitation," in *Rehabilitating Damaged Ecosystems,* J. Cairns, Jr., Ed. (Boca Raton, FL: CRC Press, 1988), pp. 105-122.
9. Cairns, J., Jr. "Quantification of Biological Integrity," in *Integrity of Water,* R. K. Ballentine, and L. J. Guarria, Eds. (Washington, D.C.: U. S. Environmental Protection Agency, Office of Water and Hazardous Materials, 1977), pp. 171-187.
10. Cairns, J., Jr. "Estimating the Assimilative Capacity of Water Ecosystems," in *The Biological Significance of Environmental Impacts,* R. K. Sharma, J. D. Buffington, and J. T. McFadden, Eds. (Springfield, VA: National Technical Information Service, 1976), pp. 173-189.
11. Cairns, J., Jr. "Aquatic Ecosystem Assimilative Capacity," *Fisheries* 2(2):5-7, 24 (1977).
12. Campbell, I. C. "A Critique of Assimilative Capacity," *J. Water Pollut. Control Fed.* 53(5):604-607 (1981).
13. Westman, W. E. "Some Basic Issues in Water Pollution Control Legislation," *Am. Sci.* 60:767 (1972).
14. Cairns, J., Jr. "Discussion of a Critique of Assimilative Capacity," *J. Water Pollut. Control Fed.* 53(11):1653-1655 (1981).
15. Ward, J. V., and J. A. Stanford, Eds. *Regulated Streams* (New York: Plenum Press, 1979).

16. Cairns, J., Jr. "Integrated Resource Management: The Challenge of the Next Ten Years," in *Aquatic Toxicology and Hazard Assessment, STP971*, W. J. Adams, G. A. Chapman, and W. G. Landis, Eds. (Philadelphia: American Society for Testing and Materials, 1988), pp. 559-566.
17. Freed, L. A., and R. L. Cann. "Integrated Conservation Strategy for Hawaiian Forest Birds," *BioScience* 39(7):475-476 (1989).
18. Kohm, K., Ed. *Balancing on the Brink: A Fifteen-Year Retrospective on the Endangered Species Act* (Island Press, in press).

CHAPTER 3

Managing Environmental Risks: Ethical Plumb Lines

Margaret N. Maxey

INTRODUCTION

Sixty years have elapsed since Alfred North Whitehead pointed out that the very nature of modern science and technology imposes on humanity the necessity for wandering, for making the transition from generation to generation "a true migration into uncharted seas of adventure." Far from being a fearsome liability to be avoided at any cost, Whitehead insists that "the very benefit of wandering is that it is dangerous and needs skills to avert evils.... It is the business of the future to be dangerous."[1]

Many have come to regard it as an ivory-towered proposition to suggest that what is dangerous might actually be beneficial to humanity. Indeed, a majority now seems to be persuaded that modern science and technology have imposed upon humanity an intolerable burden of risks and uncertainties that has not only overwhelmed our coping capacity, but has insidiously stripped us of our traditional moral moorings. Public polls reveal a near-consensus that environmental pollutants are considered not simply dangerous and harmful but morally wrong. Professor P. Sandman[2] of Rutgers University states: "Our society has reached near-consensus that pollution is morally wrong — not just harmful or dangerous . . . but wrong."

These conflicting viewpoints have a direct bearing on deliberations about the present and future management of the Savannah River itself, of the enterprises located along its site, and of the living species affected by policy decisions. The collective purpose is, first, to consider all components and interactions contributing to the complexity of environmental management, and second, to influence the thinking that will contribute to developing and implementing plans for environmental management. The outcomes of further discussion of these matters will differ significantly depending on whether, consciously or unconsciously, Whitehead's claim is embraced that future dangers call forth beneficial skills that would otherwise lie dormant, or the counterclaim is persuasive that the very existence of such dangers amounts to a moral wrong. The consequences for policy decisions are far-reaching.

If the viewpoint is conceded that the business of the future is to be dangerous, does this concession logically require the inference that what is dangerous is harmful, hence immoral? Or are dangers beneficial? How can this decision be made?

Such questions lead to a more fundamental problem: what is the business of ethics? Stated differently, what business does ethics have in the public controversy over the "morality" — not to mention the appropriate management — of "environmental risks"? Do practitioners of ethics claim to occupy some unbiased position of an impartial observer who can arbitrate public conflicts of interest or values? No one who has observed the halting, fitful emergence of both "bioethics" and "environmental ethics" as new fields of specialization will take such a claim seriously. If ethicists cannot get their own house in order and reach some consensus about norms for judging moral justification, what business do they have intruding into territory already in disarray?

Possible answers to these questions may help clarify the reasons why disagreements on environmental matters are so fundamental and why an analytic framework must be fashioned for dealing constructively with conflicts of interests and values that plague any public debate over the proper management of environmental risks. This brief essay offers only some tentative answers to these vexing questions. This discussion can only suggest how certain modes of ethical analysis might be enlisted to clarify disagreements that stand in the way of reaching acceptable policy decisions about managing environmental risks.

Possible answers to three questions are explored. First, what is the business of ethics both in general terms and in specific relation to public controversy over environmental issues? Second, how do the sources of disagreement about identifying and assessing "environmental risks" illuminate the ethical problems posed by policies for managing them? Third, what ethical plumb lines might serve as guidelines for managing environmental risks? These considerations are offered not to convince anyone to adopt a particular viewpoint, but rather to suggest ways in which a more comprehensive analytic framework might facilitate decisions about policy recommendations.

WHAT IS THE BUSINESS OF ETHICS?

Ethics today is the legacy of a collapse and fragmentation of a single moral and cosmic worldview that prevailed in Western Christendom until the Reformation. Instead of a particular religious or moral orthodoxy imposed by force, ethical reflection now transpires within secular pluralist societies marked by a tolerance for diverse moral convictions. The presumption that humans in secular societies can and should be governed by a single moral viewpoint, derived from a single supreme moral authority, has decisively disappeared from Western culture.

Attempts to reinstate some absolute standard in ethics that would compel conformity with established "moral behavior" have been exhaustively analyzed by Tristam Englehardt.[3] He examined various attempts to justify a single standard by an appeal either (1) to self-evidence in moral thought (e.g., intuitions); (2) or to the unique nature of moral reasoning (e.g., rationality or impartiality); (3) or to conformity with an external objective reality (e.g., structures of reality/nature or consequences of actions). Englehardt demonstrates how each appeal inevitably fails. In the first case, conflicts among intuitions are inescapable and consequently beg the question of some objective standard. In the second case, appeals to criteria for what constitutes reason or rationality can yield logical standards, but not moral content. In the third appeal, discovering the laws of nature or structures of reality may instruct about what is, but not what ought to be; whether man submits to structures of nature, or transforms them for human purposes, depends upon normative standards otherwise derived.[3]

Without objective normative standards and an authoritative basis for judging the morality or immorality of an Adolf Hitler or an Albert Schweitzer, man is, as Englehardt observes on page 39 of his volume, "left standing on the yawning abyss of nihilism."[3] He suggests escape from nihilism by focusing on what, in fact, at the very least, ethics is seeking — namely "a means for resolving controversies regarding proper conduct on bases other than direct or indirect appeals to force as the fundamental basis for resolution." In short, "ethics is an enterprise in controversy resolution."[3]

In modern secular societies, the vacillation and complexity of individual interests, the diversity of conflicting moral viewpoints, and the procedural constraints imposed on public policy decisions have compelled practitioners of ethics to redefine the nature, goal, and task of ethics. Englehardt proposes a redefinition that comes to terms with three matters of fact: the fragmentation of a Western world view, the necessity of reaching some general moral consensus as the foundation for public policy, and brute force is merely brute force devoid of moral authority.

The redefinition proposed by Englehardt is grounded in his search for a source of moral authority for public policy, i.e., a search for a defensible justification. When people ask ethical questions, he insists, they are searching for rational answers, i.e., for reasons why a particular controversy ought to be resolved in one way rather than another. Reasons are the antidote to coercion and brute force. It is possible, of course, that in giving reasons a particular controversy might be resolved because one party to the dispute undergoes a conversion to the other party's moral viewpoint. Or perhaps the power of intellectually sound argument might be decisive in establishing for all concerned a secular view of good as authoritative. However, history, since the collapse of the Enlightenment ideal of rational justification, attests that these options are thin reeds on which to lean for a lasting consensus.

Since pluralistic secular societies preclude expectations of conversion as well as "enlightenment," the only mode of resolution remaining as an alternative to force is procedural: namely, the free agreement of participants in a controversy to a mechanism or procedure for the peaceable negotiation of their dispute. The minimum notion of ethics as an alternative to force commits no one to a particular concrete moral view of good. Neither does it commit anyone to feign disinterest with regard to personal advantage. It construes ethics ". . . as a means for commonly and peaceably discerning or creating canons of moral probity. . . ."[3] Probity commits one to a morality of mutual respect and peaceable negotiation as a basis for the language of praise and blame about actions. Common consent is the source of moral authority for decisions to act in pursuit of particular common goods.

The moral authority acquired by public policy is not without clearly defined moral limits. Just as this authority is not derived from a deity, nor from a single moral viewpoint, neither is it derived from majority decisions. A public policy acquires moral authority only through the mutual agreement of those involved who consent to peaceable negotiations — including persuasion, inducements, and market forces. Moral limits of public authority can also be defined in terms of "natural rights" as "those that one can never presume individuals to have ceded, without threat of force, to a social organization."[3] (As examples of such natural rights, Englehardt lists "the right to use contraception, to have abortions, and to commit suicide.") These rights remain secure against majority decisions. Englehardt is emphatic: "Even a vote of all save one for their abolition does not affect their moral standing." An individual in private actions with himself or with consenting others radically limits a community's authority over that individual. Individual decisions and actions remain a private issue "unless those individuals either violate the conditions of a peaceable community generally or agree to submit their private lives to a particular community's regulation."[3] In short, the moral authority of a public policy is limited by the extent to which rights of self-determination in private matters are not ceded by coercive force and the extent to which individuals involved consent to peaceable negotiations.

The need for free mutual agreement in fashioning public policy intensifies the inescapable tension between two moral principles: (1) autonomy or self-determination, and (2) the bonds of beneficence or the provision of goods that serve the best interests of all who are affected by a social policy.

Englehardt's redefinition of the nature, goal, and task of secular pluralist ethics does not entail an abandonment either of one's moral tradition (Hindu or Judeo-Christian, Baptist, or Catholic) or of one's commitment to a single moral viewpoint. To the contrary, it entails living one's life within two moral tiers. In one tier, an individual's particular moral viewpoint can find unconstrained expression within a moral community of like-minded members. In the other tier, commitment to the moral principle of mutual respect constrains any individual from imposing a particular viewpoint on members of other communities. The two tiers are cemented by the expectation that reciprocal tolerance will prevail and that disputes will be settled by mutual agreement to resort to peaceable negotiation.

Applying the business of ethics in general to its task in relation to public controversy over environmental policy issues requires the ethical principle of mutual respect and reciprocal tolerance to be both an ethical plumb line and a moral constraint. Secular ethics do not require "conversion" to anyone's particular moral viewpoint about the presumed benefits or harms that may seem to be derived either from a natural environment or from technologies accused of degrading it. Neither does secular ethics expect an "enlightenment" to dispel the ignorance of parties involved concerning the "realities" of an external state of nature established by the preponderance of scientific truth.

The best that ethics can be expected to do in a secular pluralist world is to display reasons why common consent to peaceable negotiations for resolving environmental issues is not only morally justified but necessary as an antidote to coercion and brute force. Free agreement to negotiate peaceably is both the source of moral authority and the keystone in the arch shielding lives in a moral community.

The origins of claims and counterclaims about "environmental risks" that have contributed to deep-seated disagreements about how they should be managed can now be considered in the context of this clearly delineated secular pluralist ethic. This involves a search for reasons for resolving disputes by common consent.

"ENVIRONMENTAL RISKS": SOURCES OF PUBLIC CONTROVERSY

The past 25 years have been dominated by public controversy about the nature and magnitude of threats to the environmental quality of the entire biosphere. Rachel Carson's[4] gifted prose captured public attention in 1962 when she dramatically portrayed the devastation wrought by technological abuses of the natural environment since the 1940s. Her particular focus on the petrochemical industry and chemical pollutants quickly galvanized public protestations against indiscriminate dissemination of insecticides and herbicides labeled as "carcinogenic poisons."

There is little doubt that Carson's seminal work precipitated public debate in three major areas of ethical concern. One area of controversy is focused on claims that technological man has unprecedented power to inflict risks upon the natural environment, vs. counterclaims that far greater risks and harms are routinely inflicted upon humans and other valued species by a natural environment. Closely related is another controversial area focusing on the environmental origins of "an epidemic of cancer." A third controversial area has recently emerged as a result of claims that the "risks" targeted for regulatory concern and control are not baldly "out there" in a physical state-of-affairs, but instead are the outcome of a risk-selection process derived from social and cultural biases. A closer analysis of each of these areas of controversy reveals the source of fundamental disagreements not only about the criteria being used to identify "environmental risks," but also profound differences in methods being used to assess putative risks.

The Natural Environment As Victim

Carson's vision of a world destroyed by man's reckless and irresponsible chemical poisoning of defenseless nature has been widely accepted. However, three of her underlying assumptions have only recently been subjected to a rigorous re-examination.

Her first assumption appears in an early chapter of *Silent Spring* in which she declares that not until this century, in a singular moment of time, has one species — man — acquired through modern science and technology a magnitude of power such that it can transform not only an entire planet but the very character of power itself. In past eons, the elapse of time enabled species to adjust to change; in the modern world, the rapidity of change has overtaken time to adjust. Man's inventive mind has produced not only "synthetic creations...brewed in his laboratories, and having no counterparts in nature,"[4] but also a universal contamination of the environment. Indeed, laments Carson, ". . . chemicals are the sinister and little-recognized partners of radiation in changing the very nature of the world — the very nature of its life."[4]

Implicit in Carson's first assumption is her second. Throughout past millenia, she avers, "the environment" has rigorously shaped and directed evolving and diversifying forms of life that, given the unhurried pace of change, have reached an evolutionary plateau or "balance of nature." Carson warns against the mistake of thinking that this balance prevailed only in an earlier, simpler world. Although it is not the same today, ". . . it is still there: a complex, precise, and highly integrated system of relationships between living things which cannot safely be ignored any more than the law of gravity can be defied with impunity by a man perched on the edge of a cliff. The balance of nature is not a status quo; it is fluid, ever shifting, in a constant state of adjustment."[4] Although only theoretical, the dire consequences of upsetting nature's fragile and precarious balance with sinister synthetic chemicals appear to her to be ominous.

Within the conceptual framework suggested by Carson's first and second assumptions, it is understandable why "environmental risks" have come to be defined primarily as *risks to* a natural environment — namely, to wetlands and estuaries, to wilderness areas and scenic vistas, to ecosystems and endangered species. These risks have been identified for ethical concern because they are inflicted by biotechnological innovations driven by goals that may seem to be morally unobjectionable, i.e., increasing food production, including agriculture and aquaculture, agribusiness, and aquabusiness. In fact, their side effects are morally questionable: for example, occupational displacements; risks from pollutants in air and water, from toxic wastes, both chemical and radioactive. Ultimately, risks to a natural environment are unavoidably risks to humans and other valued species; hence such risks become morally questionable.

Carson's third assumption follows inexorably. Having reached an adjustment with past hostile elements and an ancient array of destructive forces contained in a natural environment, life must now do battle with new artificial chemical and physical agents possessing powerful capacities for inducing biological change. Carson insists that these agents do not exist in nature but have been induced by technological man, ". . . for man, alone of all forms of life, can *create* cancer-producing substances, which in medical terminology are called carcinogens."[4] With the dawn of the Industrial Revolution, Carson maintains, these non-natural agents have planted "seeds of malignancy" against which humans have no protection. In other words, Carson assumes that nature as it exists is noncarcinogenic and virtually benign. Nature poses few remaining threats to human health, since only the most resistant forms of life have survived to this point by overcoming ancient disease germs. These germs have been replaced by man-made, life-threatening carcinogens. Paradoxically, Carson does not regard her claim as a reason for despair. Because of the fact that ". . . man *has* put the vast majority of carcinogens into the environment . . . he can, if he wishes, eliminate many of them."[4] After all, most of these carcinogenic agents have only become entrenched in the world, not out of necessity but, ironically, merely to satisfy "man's search for a better and easier life."[4] Most, if not all, chemical carcinogens can be exorcised from this noncarcinogenic Nature. Carson's third assumption appears to be the fountainhead of a second major area of public controversy, namely, the environmental origins of cancer in industrialized societies.

The pervasive influence of Carson's vision of man's role in creating a cancerous universe, and of non-natural causes of cancer, has been documented by Edith Efron.[5] The second report of the Club of Rome in 1974 announced unequivocally at the outset: "The world has cancer, and the cancer is man."[6] In congressional testimony while serving as Deputy Administrator of the EPA in 1975, John Quarles clearly echoed Carson's seminal ideas:

> "Since Darwin man has recognized the ability of living things to adapt to their environments. The great diversity of life in our biosphere reflects the successful resistance of man and other species to the myriad of chemicals found in nature. However, the advent of chemical technology in the past decades has introduced billions of pounds of new chemicals that are often alien to the environment, persistent, and unknown in the interactions with living things."[7]

Less well known as a mentor and primary instructor of popular assumptions about non-natural carcinogens is Umberto Saffiotti of the National Cancer Institute. In a 1976 paper delivered at a cancer conference and designed to gain political support for passage of the Toxic Substances Control Act, Saffiotti set forth ideas based on his fundamental assumption, namely that industrial chemicals are the primary source of cancer in modern society. His view is stated unambiguously: "I consider cancer as a social disease, largely caused by external agents which are derived from our technology, conditioned by our societal lifestyle and whose control is dependent on societal actions and policies."[8] At the time he delivered his paper, Saffiotti knew that most industrial substances had never been tested for carcinogenicity. Consequently, his choice of the personal pronoun "I" signaled a personal belief; it was not, observes Efron, a scientific conclusion derived from a disciplined study of data. Saffiotti, in effect, endowed his personal belief with the status of a toxicological axiom. Indeed, he went even further. He boldly defined the toxicological policy he intended to see applied in the case of inconclusive data falling into a gray area:

> "The most 'prudent' policy is to consider all agents, for which the evidence is not clearly negative under accepted minimum conditions of observation, as if they were positive... In other words, for a *prudent toxicological policy, a chemical should be considered guilty until proven innocent*."[8] (emphasis added)

Latent in this policy is a presupposition about moral rectitude: no amount of any carcinogen may be deemed safe, since no one knows the amount required to trigger the growth of a malignancy. Moral actions require the assumption that there is no threshold dose below which exposure to a carcinogen is "safe." Even one molecule of any carcinogen must (morally) be presumed to be a potential biohazard.

Here is the moral ideal that has been transformed into a moral imperative by regulatory policy governing the protection of public health from environmental "pollutants" presumed to be the causes of cancer. A particular view of morality has dictated the quest for "zero pollution" grounded in the belief that "the only safe dose is zero dose." This view of morality underpins the criteria used by critics of modern industrial societies to identify and measure "environmental risks," i.e., primarily, their status as man-made, non-natural, synthetic creations without natural counterparts; and second, their detectability above the only safe dose of zero.

The Natural Environment As Culprit

During the past decade, the fires of controversy have been stoked by articles appearing in scientific journals, with only minimal attention paid to them by the popular media. Nonetheless, in the late 1970s, scientists began to turn their professional attention to overwhelming evidence of the existence, magnitude, and pervasive ubiquity of natural carcinogens. Efron[5] wryly remarks that "... within three years of the passage of the Toxic Substances Control Act, conceived to protect the earth from Faustian man, the earth itself had been reported to be carcinogenic beyond anyone's wildest imaginings."

In 1979, James A. Miller, a cancer researcher at the University of Wisconsin, undertook an extensive comparison of the types of carcinogens produced by nature with those produced by man. Whereas Carson told her readers that any natural carcinogens remaining were "few in number" and that humans had long ago adjusted to them, Miller[9] reached the opposite conclusion that "... a wider variety of carcinogens occurs in our natural environment.... The great majority of these agents have undoubtedly been present throughout evolution, some may even have facilitated the speciation of living systems."

In 1983, a storm of protest greeted Bruce Ames'[10] provocative article, "Dietary Carcinogens and Anticarcinogens," followed by his response to Letters in *Science*.[11] If any doubt remained, his exhaustive analysis makes it clear that humans are virtually immersed in a sea of toxic substances inherent in nature. The human diet contains a plethora of natural mutagens and carcinogens. His readers are asked to consider mushrooms laced with hydrazines, peanuts with aflatoxin, pumpkin pie with myristicin and safrole, and seafoods permeated with toxic microorganisms. Professor Ames suggests that current preoccupation with comparatively small amounts of man-made effluents from various industrial processes has diverted scientific research away from the most health- and cost-effective ways to reduce the perceived burden of cancer, namely by making dietary and lifestyle changes. In recent public testimony, Ames has declared unequivocally:

> "The main current fallacy in our approach to water pollution consists in believing that carcinogens are rare and that they are mostly man-made chemicals. Quite the contrary is the case. My estimate is that over 99.99% of the carcinogens Californians ingest are from natural (e.g. substances normally present in food) or traditional sources (e.g. cigarettes, alcohol, and chemicals formed by cooking food.)"[12]

In his reflections on the origin of spontaneous cancer, John Totter[13] has shown that mortality from cancer appears independent of the level of industrialization in a country and thus independent of its man-made pollutants, particularly when corrected for competing risks. He maintains that it is not among man-made agents that one should look for primary carcinogens, but instead among all-pervasive "natural" or "normal" environmental components. Totter suggests that the culprit is oxygen: it is a recognized mutagen; experiments have shown that it causes tumors in fruit flies; in the Ames assay tests for screening carcinogens, it shows up positive. Should regulators be morally required to ban the presence of oxygen above prescribed limits? Public controversy has more recently been fueled by critics who have called in question now-popular assumptions that nature's precarious balance is threatened by modern technologies having an enormous potential to inflict irreversible destruction of ecosystems on a planetary scale.

In the past few years, citizens in industrialized nations have grown accustomed to an almost-ritual incantation of vivid symbols of technological hazards: Love Canal, Three Mile Island, Bhopal, Chernobyl. Each is a catastrophic reminder that modern technologies have made human life more precarious and unsafe. Man seems besieged by ominous uncertainties, encroaching on life in an otherwise "safe, natural world."

Critics of this litany of technological hazards insist that the public should be confronted with a more comprehensive framework in which to assess their moral status. Each of these tragic reminders should be reviewed.

In 1978, Love Canal was the subject of a frightening media expose. Hooker Chemical was summarily tried and found guilty of harming the public with its sloppy management of chemical wastes. However, no immediate deaths occurred as a result of exposure to these wastes. To be sure, biostatisticians and epidemiologists have made terrifying predictions about hypothetical cancers and genetic defects, but they are only statistical and hypothetical.

In 1979, Three Mile Island traumatized the public. Threats to public safety from radiation exposures from accidental releases compelled the governor of Pennsylvania to order mass evacuations. No immediate deaths were ever attributed to the radiation exposures. Again, biostatisticians and epidemiologists have made terrifying predictions about *hypothetical* cancers and genetic defects.

Bhopal, in 1984, attracted not only media coverage but also a horde of trial lawyers. Some 2000 immediate deaths were caused by chemical toxins afflicting people who moved in close proximity to the plant after its construction in a sparsely populated area.

Chernobyl, in 1986, caused the immediate deaths of 31 workers who fought the fire and who received large doses of radiation. Mindboggling statistical predictions have been made of cancer deaths over the next 70 years — based entirely on the linear/zero-threshold hypothesis — with little recognition of its *hypothetical status*.

Those who define "environmental risks" in terms of the likelihood of harm inflicted by a natural environment insist that public attention should be focused on a supreme irony. In the same time frame as these so-called technological catastrophes, nature conspired to perpetrate the following events:

- In May 1985, 10,000 immediate deaths in Bangladesh were caused by a cyclone
- In September 1985, 10,000 immediate deaths in Mexico City were caused by an earthquake
- In November 1985, 20,000 immediate deaths in Armero, Columbia were caused by a volcanic eruption triggering an avalanche of mud
- In August 1986 (a mere 3 days after Soviet scientists convened an international body of colleagues to discuss the causes of the Chernobyl accident), 2000 immediate deaths at Lake Nios, Cameroon, were caused by a release of toxic gases of undetermined origin

Those who view the natural environment more as culprit than victimconcede: there are worse things in life than death. An "immediate death" in prospect appears less dreaded than a lingering death from cancer that therefore, appears more worthy of ethical argumentation concerning rights, fairness and equal protection. However, appearances are deceptive. Prolongation of life expectancy is not an unqualified good; neither is exposure to technological risk an unjustifiable harm. The point at issue is the magnitude of risks posed by nature in contrast to risks posed by human technologies.

Critics of a "benevolent nature" recall that history is replete with stark lessons about the harm inflicted on humans and their life-sustaining biosphere by the natural environment with its heedless malevolence. The Johnstown flood in 1889 caused 2209 deaths. Winds from the Great Hurricane of 1900 caused a storm surge in Galveston, TX that claimed 6000 lives. A volcanic eruption at Mont Pelee, Martinique, in 1902 caused 30,000 immediate deaths. These events surely qualify as "hazard spectaculars" against which human beings have tried to defend themselves. Engineers have devised systems of flood control, as well as early warning systems for hurricanes and volcanic eruptions.

The tragic reminders of the destructive power of nature serve not to exonerate human fallibility nor to excuse technological failures. Instead, they heighten a sense of man's precarious achievements in technological control over nature's destructive forces. It is specious to argue that since humans cannot prevent or control the forces of nature, man ought to concentrate time, effort, and public money on eliminating technological risks. To the contrary, technologies throughout history have been remarkably successful in protecting human life from exposure to nature's harmful effects.

These claims and counterclaims about *risks to* vs. *risks from* a natural environment offer ample reasons to conclude that neither the truth nor the falsity of either one of these sets of claims be conceded as an *ethical basis* for resolving disputes about "managing environmental risks." Each side has a persuasive case — of course, with divergent policy implications — that proper and appropriate management is not simply an option for decision makers, but a necessity.

The unresolved question remains: what will count among disputants as "appropriate management?" This question requires the examination of the third area of controversy.

Environmental Risks: Inflicted or Selected?

The third major area of public controversy has only recently emerged. Among influential writers who first introduced the dispute are William Clark[14,15] and Mary Douglas and Aaron Wildavsky.[16]

To William Clark, neither risks nor hazards graphically exist "out there" in nature or in human transactions with it. What people regard as hazardous in any given era reflects what they have come to know about their environment and what they value as essential or desirable on a scale of real possibilities. In short, human beings structure hazards. In that sense, hazards are human artifacts. A hazard is not by definition "toxicity of substance" or "violence of event" or "magnitude of consequences" that can be known, classified, and predicted. A hazard exists only when, and to the degree that harmful exposure of and assimilation by the human body or other valued living systems become a genuine, not merely a hypothetical, possibility. That possibility exists only when there is a management failure to devise and maintain controlling actions or safeguards. According to Clark, hazard management is "the adaptive design of hazard structure," and the primary goal of hazard management is "to increase our ability to tolerate error and to take productive risks."[14] The ultimate hazard is the failure to design and maintain structures of social resilience.

To Douglas and Wildavsky, the primary question is how to account for a puzzling anomaly. In less than a generation, affluent citizens have been transformed into fear-ridden critics who protest against technological hazards, environmental pollution, and personal contamination caused by "corporate cancer." Why have science and technology, regarded for centuries as preeminent sources of safety and reliable instruments of human control over hostile nature, become the object of skepticism and doubt?

Sociologists and psychologists customarily attribute differing perceptions of risk to different personality traits. However, Douglas and Wildavsky maintain that risks are selected for attention dependent upon the strength of social criticism and cultural bias. People pay attention to selected risks, not arbitrarily but purposefully, i.e., to reinforce and conform to a specific way of life already selected on other grounds. When people disagree about risks that are merely "perceived" vs. "actual" risks they may be disposed to take or to avoid the disagreement signals divergent agendas for changing or preserving preferred forms of social organization.

Persuasive reasons exist for the view that current disputes over the discrepancy between actual vs. perceived risks cannot be accounted for if the premise is accepted that real knowledge of an external world should be allocated to experts in the physical sciences, while mistaken perceptions pertain to the realm of personal psychology. According to this faulty division, dangers are assumed to be inherent in a physical state-of-affairs, and the risks they pose are objectively self-evident and ascertainable by experts. Subjective personality traits, i.e., an individual is either a risk taker or a risk avoider, cannot account for a bias toward the selection of certain types of risk and not others. A subjectivist view can be nullified by two questions: Why is it that experts disagree? Why does an individual fear environmental risks to the exclusion of those dangers, e.g., smoking or speeding on an expressway or overeating, that are more immediate threats to life expectancy?

Douglas and Wildavsky insist that only a cultural theory of risk selection as a product of cultural bias and social criticism can account for the anomalies humans encounter in affluent technological societies: "Only a cultural approach can integrate moral judgments about how to live with empirical judgments about what the world is like."[16] The risks selected for control or mitigation, individually and collectively, are integral to the choices made with respect to the best way to organize social relations, to protect shared values, and to devise institutional mechanisms for providing informed consent in fashioning public policy.

In the current dispute over environmental risks and technological hazards, partisans accuse each other of belonging either to "the danger establishment" or to the "industrial establishment." Each accuses the other of serving the vested interests of preferred social institutions. Each is accused of irrational bias, of misperceptions of *real* risks, and of subversion of the public interest.

The outcome of this controversy is far from clear. However, there is little doubt that it has far-reaching implications for policy decisions about environmental risk management.

TOWARD RECONSTRUCTING RISK MANAGEMENT: ETHICAL PLUMB LINES

The previous discussion concerning the three major areas of controversy in identifying and assessing environmental risks has given reasons why the ethical problems posed by current methods of risk management intensify the tension between individual needs for autonomy and self determination straining against collective needs for social beneficence in policy decisions. These problems cannot be resolved by continuing to expect "conversions" or "enlightenment." It is an exercise in futility to continue relying on an appeal to moral intuitions about self-evidence, or to objective scientific facts, or to the impartial authority of experts as convincing tools for risk management. Such appeals appear to be secular substitutes for a displaced deity. The alternative is free agreement to negotiate peaceably.

Conflicting Concepts of Risk

Two major factors contributing to the ethical problems that inhibit common consent have been (1) the reification of the concept of risk as a thing "out there" in nature, and (2) the domination of risk assessment by technical discourse and engineering methods of analysis.

In the first instance, the dominant concept of "risk" has become defined in terms of the probability of an adverse event multiplied by the magnitude of its consequences. This concept reduces the risk management problem to one of finding methods of convincing a recalcitrant public that expert assurances about low probabilities require acceptance, or at least acquiescence. Such reductionism gives every appearance of ignoring, and even trivializing, the ethical concerns manifested in the values, expressed and implied, represented by the decisions of ordinary people. As a result of their research, Steve Rayner and Robin Cantor have concluded that, rather than probabilities multiplied by consequences, what ordinary people care about are trustworthy institutions, equity in compensation for impairments to personal health, and adequate procedures for assuring informed consent to technological choices. Consequently, they insist that the question appropriate to risk management is not "How safe is safe enough?" but "How *fair* is safe enough?"[17]

In the second instance, professionals in engineering have been influential in developing four methods of risk assessment reasoning devised to allay public fears and win acceptance of risks posed by technological innovations. One of the first methods developed comparisons of technological risks with other routine risks commonly accepted by the public in daily life, e.g., driving automobiles, smoking cigarettes, failing to install smoke detectors, refusing to wear seatbelts. Critics of these comparisons were quick to discredit them by claiming that such choices entailed voluntary risk taking, whereas a coal plant or toxic waste dump located "in my backyard" imposes on the individual an involuntary risk that is not at all acceptable. An alternative method of reasoning has insisted that proper risk assessment requires comparison with the benefits derived from accepting the risks. However, detractors of this comparison have argued vigorously that benefits exist only "in the eye of the beholder;" benefits are notoriously difficult to measure and are not evenly distributed among those who must bear the risks.

Confronted with these objections, more sophisticated methods of comparison have been developed based on comparisons among risks of alternative ways of achieving the same objective. Herbert Inhaber[18] has published a classic study on electricity, demonstrating this method by analyzing every conventional and nonconventional way of generating electricity "from cradle to grave." Critics have insisted that such risk comparisons beg the question of why new methods of generating electricity should be acceptable since they impose additional risks that are in fact unnecessary.

Still another method has emerged of comparing alleged risks of new technologies (in particular, radiation exposure from nuclear reactors) with exposures from natural background radiation. People receive wide variations in exposure to naturally occurring radioactive substances, cosmic and terrestrial. Hence, it seems eminently logical to argue that standards in effect for 50 years have set limits on technology-enhanced exposures that are well within the range of variation from nature itself. Consequently, such limits should be acceptable. Critics again lament that man-made sources simply add to a current risk background without justification.

As detractors have attacked each of these proposed methods for risk assessment reasoning in turn, it has become clear that there is a twofold objection common to and underlying each attack, emphasizing the contrast between a concept of "risk" held by ordinary citizens and that of scientists and engineers. By reifying risk as a thing "out there," a technical concept feeds perceptions of risk as an *increment,* as a simple addition to a current risk background; moreover, the "new risk" imposed by today's technologies seems to be utterly unique and incommensurable with past risks, suggesting the implication that risk reduction options cannot achieve "equal protection" by a more economically efficient use of dollars-per-life-saved.

Counterarguments to these objections have enabled participants in the controversy to appreciate the misunderstandings and misinterpretations that have been generated by the dominance of technical discourse and engineering preoccupations in the language of risk analysis. Now that social historians and humanists have entered the fray, persuasive analysis has emerged demonstrating that objections rest on the fallacy of misplaced concreteness. Any "new risk" reorders an entire system by displacing, offsetting, substituting for, or otherwise restructuring a prior pattern of benefits and harms. The concerns of ordinary people require a systemic risk assessment that surveys a total spectrum of threats to health and then compares risk-reduction options and associated costs in order to achieve the most equitable health protection for an entire populace. As for the "incommensurability" objection, scientists respond that it rests on an honest misunderstanding. Admittedly, risks are not equally distributed in a population, but this does not justify the claim that "efficient use" of risk-reduction dollars cannot achieve equity. The fallacy is that a concept of "efficiency" or avoiding wastefulness has been mistakenly substituted for the concept of effectiveness in achieving an intended objective, in this case, the protection of health from actual rather then merely hypothetical harm. Cost-effective analysis (CEA) must be carefully distinguished from cost-benefit analysis (CBA) developed in early risk assessment reasoning.[19]

Ethical Plumb Lines

A consideration of the diverse interests to be managed along the Savannah River site should focus on at least three ethical plumb lines consistent with Englehardt's redefinition of the task of a secular pluralist ethics.

In the realm of **micro-ethics**, where the focus is on the individual and relations between individuals, the principle of mutual respect endeavors to assure reciprocal tolerance for a wide range of diverse moral viewpoints on management priorities. The success of integrated environmental management will be largely dependent upon the extent to which individuals care about sustaining their resources.

The realm of **molar-ethics** focuses upon institutions as they function both internally and in relation to society as a whole. This is the realm of Englehardt's second tier in which most of the major ethical issues of environmental risk management are encountered. The ethical principle of beneficence, which requires individuals and institutions to serve the best interests of all who are affected by a policy, is clearly in opposition to autonomous individuals claiming rights to self-determination, protection from harm, and compensation for injury.

The immediate needs for reconstruction and reform of risk management occur in the realm of molar-ethics. The most insightful and urgent recommendations to date are those of Peter Huber.[20] The shifts in tort law from private risks to public risks, driven by the torrent of risk information available and uncorrected by evidence that new technologies are generally much safer than those they displace, have produced flagrant violations of the principle of equity or fairness. The new tort system ". . . delegates risk selection to lawyers, judges, and lay jurors who are magnificently unqualified to distinguish good public risks from bad ones."[20] Public risks certainly require public control. The ethical question is who is qualified to manage public risks? Huber's response distinguishes between two spheres of institutional competence. When they address private risks, i.e., "high-probability bilateral hazards that have ripened (or are about to ripen) into concrete injuries," courts and the tort system perform adequately. In the case of public risks that are "diffuse, low-probability, multilateral, and temporarily remote" and that entail risks averted as well as incurred and persons helped as well as harmed, only a public viewpoint supplied by public agencies will suffice. Huber's proposal is modest: "The courts should defer to the public-risk choices made by experts in the regulatory agencies."[21] Current law must be changed to prevent judges from overriding prior administrative determinations that public risks constitute progressive choices. Moreover, he insists that judicial management of compensation systems must be supplanted by two administrative compensation systems already in place and by offering models for more effective and equitable decision.

The third ethical plumb line pertains to the realm of **macro-ethics** where conflicting worldviews provide the ultimate grounding for public controversy over public policy. In this realm the predominant principle must be reciprocal tolerance. Ever since the biblical book of *Genesis* with its concept of "original sin," an age-old tension has pitted catastrophists or entropists against cornucopians or agents of technological substitution. Because it is the business of the future to be dangerous, this tension will be forever with us, and that is all to the good, for without it, life would be a barren horizon of complacency. As for those who would eliminate this tension, in the sage words of Stephen Minot, "For the prophets of doom, only doom will suffice."

REFERENCES

1. Whitehead, A. N. *Science in the Modern World* (New York: Mentor Books, 1949) pp. 207-208.
2. Sandman, P. Rutgers University. Cited by Geraldine Cox in "The Dangerous Myth of a Risk-Free Society," public address delivered at Drexel University, Philadelphia, 10 February 1989. p. 3.
3. Englehardt, H. T., Jr. *The Foundations of Bioethics* (New York: Oxford University Press, 1986).
4. Carson, R. *Silent Spring* (Greenwich, CT: Fawcett, 1962).
5. Efron, E. *The Apocalyptics: Cancer and the Big Lie* (New York: Simon & Schuster, 1984).
6. Gregg, A. "A Medical Aspect of the Population Problem," *Science* 121:681 (1950).
7. Quarles, J. In *The Apocalyptics: Cancer and the Big Lie,* E. Efron, Ed. (New York: Simon & Schuster, 1984), p. 125.
8. Saffiotti, U. "Risk Benefit Considerations in Public Policy on Environmental Carcinogenesis," in *Proceedings of the 11th Canadian Cancer Research Conference* (Toronto: National Cancer Institute, 1976).
9. Miller, J. A. "Concluding Remarks on Chemicals and Chemical Carcinogenesis," in *Carcinogens: Identification and Mechanism of Action,* A. C. Griffin, and C. R. Shaw, Eds. (New York: Raven Press, 1979).
10. Ames, B. "Dietary Carcinogens and Anticarcinogens," *Science* 221:1256-1264 (1983).
11. Ames, B. Reply to "Letters," *Science* 224:760 (1984).
12. Ames, B. "Water Pollution, Pesticide Residues, and Cancer." Adapted from November 11, 1985, testimony, Senate Committee on Toxics and Public Safety Management. *Water* 27(2):23-24 (1985).
13. Totter, J. R. "Spontaneous Cancer and Its Possible Relationship to Oxygen Metabolism," *Proc. Natl. Acad. Sci. U.S.A.* (1980).
14. Clark, W. C. "Managing the Unknown," in *Managing Technological Hazard: Research Needs and Opportunities,* R. M. Kates, Ed. (Boulder: University of Colorado Institute of Behavioral Science, 1977).
15. Clark, W. C. "Witches, Floods, and Wonder Drugs: Historical Perspectives on Risk Management," in *Societal Risk Assessment: How Safe is Safe Enough?* R. C. Schwing, and W. A. Albers, Jr., Eds. (New York: Plenum Press, 1980), pp. 284-318.
16. Douglas, M., and A. Wildavsky. *Risk and Culture: An Essay on the Selection of Technological and Environmental Dangers* (Berkeley, CA: University of California Press, 1982).
17. Rayner, S., and R. Cantor. "How Fair is Safe Enough? The Cultural Approach to Societal Technology Choice," *Risk Anal.* 7(1):3-9 (1987).

18. Inhaber, H. *Energy Risk Assessment* (New York: Gordon & Breach, 1982).
19. Emery, D. D., and L. J. Schneiderman. "Cost-Effectiveness Analysis in Health Care," *Hastings Center Rep.* 19(4):8-13 (1989).
20. Huber, P. W. "Private Risk and Public Risk," *Chemtech* 18(8):467-470 (1988).
21. Huber, P. W. *Liability: The Legal Revolution and Its Consequences* (New York: Basic Books, 1988).

CHAPTER 4

The Systems Approach to Environmental Assessment

R. V. O'Neill

INTRODUCTION

The systems approach in ecology is both a methodology and a holistic mindset. As a methodology, it utilizes mathematical and computer models to simulate natural ecosystems. As a mindset, it focuses on the total functioning and overall properties of the ecosystem.

The distinction between methodology and mindset forms a convenient outline for this discussion. Under the rubric of methodology, the applicability of ecological models to environmental assessment are discussed, and under the auspices of the holistic mindset, the feasibility of assessing impacts on overall ecosystem function is considered.

In both cases, the systems approach has much to offer in assessing environmental impact. The approach encourages a comprehensive viewpoint and avoids the pitfalls that result from focusing too closely on a single population or process. At the same time, the systems approach is no panacea. No single approach should be relied upon (or ignored) in addressing complex assessment issues.

SYSTEMS ANALYSIS AS METHODOLOGY

A number of book-length treatments are available that provide access to mathematical and computer tools on modeling aquatic ecosystems,[1-3] on applied problems,[2,4,5] wildlife management,[6] fish populations,[7] ecotoxicology,[8] and aquatic ecosystems.[9,10] These volumes are backed by significant journal literature on the use of systems analysis in assessment. Recent studies have applied the approach to water quality,[11,12] harvesting marine populations,[13] controlling insect pests,[14-16] air pollution,[17,18] and crop management.[19]

This literature indicates that systems analysis and modeling has become an integral part of environmental science. A similar impression is gained by surveying recent volumes (1986 to 1989) of major ecological journals. In *American Naturalist*, 49% of the articles contain mathematical models. In *Ecology*, 21% of the articles contain models at some level of resolution. If the development of ecology follows other natural sciences, the increasing use of mathematical tools can be taken as a positive sign of progress and maturation.

The increasing use of systems analysis methodology does not necessarily imply that ecological models always provide accurate predictions. If modeling objectives are well focused and critical scales and processes are understood, ecological models can be successful. An excellent example is provided by models of succession following the opening of a gap in a forest canopy.[20] The models are limited in scope, scaled at the level of the individual tree, and rely heavily on well-understood processes of competition for light. As a result, the models tend to compare well with independent data.

In many instances, however, significant gaps occur in understanding ecological processes. In general, framing ignorance in equations does nothing to overcome the ignorance. The model is limited to the insights and understanding that have been supplied by the researcher.

In other instances, problem objectives are not well defined. In these cases, model success or failure is difficult or impossible to judge. Problem definition must necessarily include clear criteria for evaluating model performance.

The real assets of the systems analysis approach can be summarized in two words: synthesis and consequences. The modeling format permits synthesization of an enormous amount of ecological data and insight into a manageable tool. Then, the tool will slavishly work out the logical consequences of this understanding. Most importantly, the model can deal with potential indirect effects that are largely inaccessible using other approaches.

If the understanding and data are inadequate to address assessment objectives, model predictions will not be correct. However, the lack of adequate understanding should be blamed, not the methodology. The model is simply (1) synthesizing and (2) drawing out the implications of current understanding. When interactions become very complex, modeling is about the only tool for augmenting mental faculties and examining implications. The model, correctly scaled and formulated, is at least a faithful servant of ignorance.

One constructive approach to evaluating model uncertainties is risk analysis.[21] In this approach, the model is run across a range of possible parameter values and/or mechanisms. The results are stated in terms of the probability or risk that a given action will result in an unacceptable effect. Particularly interesting is the finding that certain patterns of toxicity across food chains are likely to result in undesirable risks.[22]

To date, the applications of risk analysis have emphasized uncertainties in measured data rather than uncertainties in ecological mechanisms. The approach offers little to resolve the general problem of ignorance since an unknown mechanism will always have an unknown effect.

In spite of the ignorance, probabilistic approaches to prediction can be of real value in decision making. Consider, for example, the problem of predicting today's weather. The physical mechanisms determining local weather patterns are not well understood, but through a combination of models and accumulated past experience, a probabilistic prediction is possible, i.e., it is possible to state the probability of rain today. In most cases, this information makes a valuable contribution to the complex decision process that determines whether a person will carry an umbrella.

In a similar manner, the current generation of ecosystem models could be used in a program of environmental "weather prediction." As a suite of models is applied over a series of assessment problems, experience would be accumulated in the success or failure of the model. In a manner directly analogous to weather prediction, sufficient experience with the model would eventually be accumulated to "forecast" the risk of undesirable effects.

Another innovative approach to the problem is analogous to the "miners' canary." A canary was always kept in a mine while workers were present. The canary was hypersensitive to toxic gases, and when it stopped singing and died, it was time to leave the mine. Similarly, a hypersensitive assessment model might be constructed. The model would include known and hypothetical mechanisms that tend to amplify environmental effects. Then, if the model predicted no undesirable effects, the conclusion could be drawn that current ecological understanding and data do not indicate a serious problem.

Of course, the suggested approach is no better than the canary. The canary does not indicate when the roof of the mine would cave in. Neither will the model warn of effects from mechanisms not included in the model. However, just because the canary did not warn the miner of all possible dangers is no excuse for forgetting to bring the bird along.

Proper scaling of the model may determine its applicability to an assessment problem. Environmental systems present a complex mosaic of spatial and temporal scales.[23] Focus on ecological processes can be at the level of the individual organism, the population, an intact system (such as a forest), or even the entire biosphere. No single scale is best or most important. Failure to select the proper scale explains many failures of assessment in general and assessment models in particular.

Perhaps the classic problem resulting from improper choice of scale is the "Smokey the Bear" campaign. When viewed at the scale of the individual tree or the forest stand, it seemed logical to rail against forest fires. Preventing forest fires saved trees. Every small fire that was extinguished saved the forest from larger and more devastating fires.

However, the impeccable logic failed to produce the desired effect. At larger spatial scales, the forest is a mosaic of stands. In the absence of fire management by European man, lightning and occasional cooking fires produced frequent fires. Fuel loads remained low, and fires were generally small and cool. The resulting landscape was diverse, with patches in many stages of recovery. When a fire burned a small patch, it was seldom many miles from a seed source, and recovery was rapid.

Fire prevention programs interrupted the natural pattern, and landscapes proceeded to a uniform cover of mature trees. Fuel accumulated as branches, and dead trees piled up on the forest floor. Then, when the inevitable fires did occur, they ranged over thousands of acres. Often the accumulated fuel made the fires hot enough to burn organic matter on the forest floor, leaving bare ground. No seed sources existed to facilitate recovery, and erosion began to strip the soil.

Attempts to control fire at the small stand level resulted in far greater fire damage to the forests at the larger scales that characterized entire landscapes. The management strategy plainly and simply missed the point that fire was a natural occurrence and could not even be regarded as a disturbance at the larger scales. By changing the size, intensity, and frequency of the recurrence of fire, the policy had disrupted the mechanisms that maintained stability at the landscape scale.

The problem of choosing the proper scale carries over directly to the problem of assessing societal effects on the environment. An excellent example of an inappropriate assessment scale is provided by the HERMES model.[24] In 1970, the United States Atomic Energy Commission authorized a study of the long-term impact of the nuclear power industry across the upper midwestern region. The model was to consider the eventual construction of 185 power plants and calculate doses to man in the year 2000.

The approach can be exemplified by the submodel for radionuclide concentrations in beef products. The entire region was modeled at the scale of the individual pasture. Small amounts of radionuclide in stack emissions were diffused downwind, settled on vegetation or were taken up from the soil, and were concentrated in beef and milk products. In other words, the region was decomposed to units of perhaps 100 acres or less, and processes within these units were modeled in some detail.

However, at the scale of the entire region and over a period of 30 years, the processes within a pasture became totally irrelevant. Economic pressures significantly altered the spatial configuration of pastures. Agricultural practices shifted from pasture feeding to feedlots. Sociological changes altered dietary intakes. Since these larger scale processes were not explicitly considered in the model, it was unlikely that the model could ever achieve its objectives.

The problem of scale can be compounded by ambiguities in the value systems that society places on the environment. Conflicting societal values may insist on preserving the aesthetic values of large regional systems while maximizing the production of a single species within this system. Often, these management objectives are not compatible.

An excellent example of conflicting values at different scales is provided by the controversy over the Indian Point power plants on the Hudson River.[7] Large intakes of cooling water from the river had the potential of impinging the young of the striped bass that spawn upriver. Early evaluations indicated that impact at the scale of the intact river system would be slight. It was likely that the major effect would be a replacement of striped bass by white perch. Overall, the system would respond stably to the imposed stress.

However, such a conclusion was unsatisfactory to the members of the Isaac Walton League, whose value system focused on sport fishing for the striped bass. The result was years of litigation over the environmental impact of the power plants. Confusion is a natural outcome of a given situation that might be characterized as catastrophic, or as a mere statistical ripple, depending on spatial and temporal scales.

In many cases, the conclusion of the inability to predict the impact of a societal action may simply be a confusion of scales. This is exemplified in a detailed validation study on Lake Keowee[25] that compared an aquatic ecosystem model with an independent data set for a reservoir ecosystem. The model failed to duplicate the field data, but the reason for the failure was not any intrinsic lack of predictability but with incompatibility of scales between the model and the data set. The temporal resolution of the data set, e.g., monthly or bimonthly sampling of phytoplankton, made it incompatible with the daily dynamics of the model. The model assumed spatial homogeneity, while the data set was from a lake with pronounced heterogeneity, depth stratification, and patchiness. The model failed to predict because the ecological insight, as reflected in the mechanisms of the model, was simply at the wrong scale for the problem being addressed.

Any serious challenge to the environmental predicament must involve a more careful consideration of scale. The scale of perturbation, i.e., its extent in space and its frequency and duration in time, is critical, and the spatial extent of an environmental system may determine whether it will respond stably to a disturbance.[26] Thus, large forests are relatively stable with respect to small fires, while a small Caribbean cay may never reach equilibrium in the face of frequent hurricanes.

Similarly, the chaparral community in Southern California is fire-adapted and recovers from fires that recur at intervals of 5 to 10 years. However, if the frequency is altered so that fire recurs on the same plot for several succeeding years, succession takes the plot to a different community of species.[27]

Modeling approaches to assessment can be summarized in three points. First, ecosystem models summarize and draw out the implications of the present understanding and data. The models are as good as, and never better than, the insights of the investigator. Second, even with limitations, imperfect models can provide useful probabilistic determinations of risk that can be integrated into the total decision making process. Third, a careful consideration of scale will avoid many pitfalls associated with past failures in applying assessment models.

THE SYSTEMS APPROACH AS HOLISTIC MINDSET

When assessment issues are focused on a critical population or single end point, the problem is greatly simplified. Current levels of understanding of physiology, toxicology, and population ecology are immediately applicable. The paradigms that bear on the problem are well defined and broadly accepted. As a result, population effects will continue to be an important practical approach to dealing with assesment problems.

On the other hand, fundamental ecological issues are left unanswered by any population-focused approach.[28] The ecosystem is a biogeochemical system[23] with a complex set of feedbacks and controls. Impacts on a single species may have negligible effects on overall system function since the affected species may simply be replaced by a competitor. For example, O'Neill and Giddings[29] argue that considerable shifts in phytoplankton communities can occur in nutrient limited systems without a detectable change in total production. On the other hand, effects on critical ecosystem processes, such as remineralization and recycling, might be detectable without knowing which of the myriad microbial populations is most sensitive to a new impact.[30,31]

Any focus on ecosystem effects, however, is hampered by a lack of methods. Potential approaches to measuring overall ecosystem function have been reviewed.[32] Any review depends heavily on previous attempts to develop test protocols for toxic effects on ecosystems.[28,33-35]

All of the methods view the ecosystem as an energy-processing system[36] that is limited by critical nutrients.[37] Both energy processing and nutrient recycling are accomplished through interactions among components. These interactions confer a degree of homeostatic control that maintains the system in the face of environmental fluctuations.[36,38] Any perturbation in these control mechanisms has immediate implications for the stability of the system. Consequently, measures of overall ecosystem function should be associated with these homeostatic mechanisms, i.e., energy flow and nutrient cycling mechanisms.

Measures Based on Energy Flow

In a stable ecosystem, the ratio of gross primary production to respiration (P/R) should remain close to 1.0.[39] Ratios approximating 1.0 have been found in many experimental[40-45] and natural ecosystems.[46-49]

Microcosm studies consistently demonstrate that P/R departs from 1.0 when a system is disturbed. This effect can be seen with temperature stress,[40] reduced light intensity,[42] or increased grazing pressure.[41,50] Various toxic substances produce the same result.[44,51] These studies indicate that changes in P/R could be a reliable indicator of stress.

Another metabolic parameter that might be of interest is power, defined as the total energy flow through the system per unit biomass. Odum and Pinkerton[52] introduced the concept and argued that greater power meant greater capability to respond to environmental insults. O'Neill[53] used simplified ecosystem models to show that power was related to the relative ability of ecosystems to recover from perturbation. DeAngelis[54] demonstrated analytically that there was a relationship between power and the time for recovery.

A sensitive measure of ecosystem function might be based on frequency response analysis. Analysis of an extensive time series of ecosystem metabolism[31] or biomass[55,56] extracts periodicities characteristic of system function. Van Voris et al.[31] proposed that the number of periodicities reflected the functional complexity of the system. Dwyer and Perez[57] confirmed this hypothesis experimentally.

At present, the use of characteristic frequencies as an indicator of ecosystem impact is limited. The approach requires extensive data sets, and changes in periodicities are difficult to relate to changes in the ecosystem. Nevertheless, the implication that the number of characteristic frequencies is related to functional complexity[31,57,58] means that the approach has valuable potential.

Measures Based on Nutrient Processing

In aquatic ecosystems, sediments may be the best location to observe alterations in nutrient cycling.[44] Close coupling between supply rates and uptake indicates that dissolved inorganic nutrients in the water column may be relatively insensitive to change.[59] The exchange between sediments and the water column controls chemical conditions in lakes and marine environments.[60-64] Processes occurring at the sediment-water interface may be measured in extracted sediment cores maintained at ambient temperatures, with adequate aeration and mixing.[65,66]

The proposed measure would involve the breakdown of complex organic compounds. Decomposition of complex compounds involves a series of populations, each operating on the product of previous breakdowns. Thus, the hypothesis is that a toxic effect on any of the populations would slow the overall decomposition rate. If, as Domsch[67] has hypothesized, organisms that degrade resistant substances are sensitive to toxic effects, changes in rates of decomposition could be a sensitive measure. In fact, good evidence exists for the sensitivity of decomposition processes to heavy metals.[45,68-74]

Measures of overall ecosystem function may be devised by focusing on the intact ecosystem as a biochemical processing unit. Baas Becking et al.[75] used two variables from chemical equilibrium analysis:[76] pH and Eh as an easily measured approximation for pE (-log electron activity). Schindler et al.[77] used a similar approach with pH and dissolved oxygen as an approximation for pE. These two variables are integrated measures of many chemical processes, and changes of a microcosm from one organizational state to the other could be related to metabolism and processing of essential nutrients. This approach is useful[78] in measuring total system response to perturbations.

In lotic ecosystems, retention of nutrients must compete with transport processes that move nutrients downhill. Webster coined the term "spiralling" for the spatially distributed cycling of nutrients.[79-81] Spiralling length[11,82-86] is the average distance traveled by a nutrient as it completes one cycle, i.e., travels from one functional component of the ecosystem back to the same component downstream. Spiralling length can be approximated as the ratio of the flux of a nutrient in the water to the rate of uptake of nutrients from the water column. Any toxic effect on rate processes in the system should be reflected, therefore, in a change in the spiralling length. The feasibility of using spiralling length in natural ecosystems has been demonstrated with a radiophosphorus tracer.[87]

Loss of limiting nutrients from terrestrial ecosystems by leaching can be important to ecosystem maintenance and productivity.[88] Extensive perturbations, such as clearcutting, can cause the loss of nutrients from watersheds.[89] Nutrient leaching has also been proposed as a sensitive indicator of ecosystem stress.[30] Research on leaching from terrestrial soil cores[45,90-92] lends support to this concept.

These data suggest three criteria for a desirable measure of total ecosystem function. First, the measure should involve one of the two basic ecosystem functions: energy flow and nutrient cycling. Second, the measured process should be the result of many interacting populations or components. In this way, impact on any one or any combination of components, might be reflected in the measurement. Third, the measure should be directly interpretable in terms of ecosystem function. Thus, the component that played a major role in reducing P/R may not be identified, but whatever happened, it placed the system into a negative energy balance.

It should be noted that measures of total ecosystem function supplement, but do not replace, population and community indicators. Ecosystem indices are designed to measure higher order effects that might be missed by population measurements. At the same time, natural compensatory mechanisms, such as species replacement, might result in no change in an ecosystem indicator even though significant changes in populations occurred. Therefore, both population and ecosystem measurements are needed to ensure the integrity of the natural system.

CONCLUSIONS

Whether the systems approach is considered as methodology or mindset, it appears that it can continue to play a constructive role in environmental assessment. Models are excellent tools for synthesizing ecological data and drawing out the implications of complex interactions, but the tool must be used with due consideration of scale and the limitations of current understanding.

Holistic measures of ecosystem metabolism and nutrient processing continue to suggest themselves as sensitive, "early warning" indicators of ecosystem damage. Continued research on total system measures could produce reliable, inexpensive approaches to environmental monitoring.

Perhaps the summary of this review is the suggestion that environmental assessment is a difficult and complex process. The systems approach presents a set of tools that should continue to be applicable to assessment. However, having an excellent hammer does not guarantee a beautiful house will be built. Judicious application and continued vigorous research are needed to ensure that the tools are properly applied.

ACKNOWLEDGMENTS

Research was supported by the Office of Health and Environmental Research, U.S. Department of Energy, under Contract No. DE-AC05-84OR21400 with Martin Marietta Energy Systems, Inc.

REFERENCES

1. Straskraba, M., and A. H. Gnauck. *Freshwater Ecosystems: Modelling and Simulation* (New York: Elsevier, 1985).
2. Walters, C. *Adaptive Management of Renewable Resources* (New York: Macmillan, 1986).
3. Swartzman, G. L., and S. P. Kaluzny. *Ecological Simulation Primer* (New York: Macmillan, 1987).
4. Frankiel, F. N., and D. W. Goodall, Eds. *Simulation Modeling of Environmental Problems* (Chichester, England: John Wiley & Sons, 1978).
5. Beyer, J. E. *Aquatic Ecosystems: An Operational Research Approach* (Seattle: University of Washington Press, 1981).
6. Grant, W. E. *Systems Analysis and Simulation in Wildlife and Fisheries Sciences* (New York: John Wiley & Sons, 1986).
7. Van Winkle, W. *Assessing the Effects of Power-Plant-Induced Mortality on Fish Populations* (New York: Pergamon Press, 1977).
8. Levin, S. A., and K. D. Kimball. "New Perspectives in Ecotoxicology," Ecosystem Research Center Report No. 14a, Cornell University, Ithaca, NY (1983).
9. Fontaine, T. D., and S. M. Bartell. *Dynamics of Lotic Ecosystems* (Ann Arbor, MI: Ann Arbor Science Publishers, 1983).
10. Reckhow, K. H., and S. C. Chapra. *Engineering Approaches for Lake Management, Vol. 1: Data Analysis and Empirical Modeling* (Ann Arbor, MI: Ann Arbor Science Publishers, 1983).
11. Dejak, C., and G. Pecenik, Eds. "Mathematical Modelling of Eutrophication and Dispersion in the Lagoon of Venice," *Ecol. Model.* 43:1-130 (1987).
12. Riley, M. J., and H. G. Stefan. "MINLAKE: A Dynamic Lake Water Quality Simulation Model," *Ecol. Model.* 43:155-182 (1988).
13. Grant, W. E., J. H. Matis, and W. Miller. "Forecasting Commercial Harvest of Marine Shrimp Using a Markov Chain Model," *Ecol. Model.* 43:183-194 (1988).
14. Byrne, S. V., M. M. Wehrle, M. A. Keller, and J. F. Reynolds. "Impact of Gypsy Moth Infestation on Forest Succession in the North Carolina Piedmont: A Simulation Study," *Ecol. Model.* 35:63-84 (1987).
15. Knudsen, G. R., and G. W. Hudler. "Use of a Computer Simulation Model to Evaluate a Plant Disease Biocontrol Agent," *Ecol. Model.* 35:45-62 (1987).
16. Longstaff, B. C. "Temperature Manipulation and the Management of Insecticide Resistance in Stored Grain Pests: A Simulation Study for the Rice Weevil, *Sitophilus oryzae*," *Ecol. Model.* 43:303-314 (1988).
17. Chen, C. W., J. D. Dean, S. A. Gherini, and R. A. Goldstein. "Acid Rain Model - Hydrologic Module," *J. Environ. Eng.* 108:455-472 (1982).
18. West, D. C., S. B. McLaughlin, and H. H. Shugart. "Simulated Forest Response to Chronic Air Pollution Stress," *J. Environ. Qual.* 9:43-49 (1980).

19. Lemmon, H. "Comax: An Expert System for Cotton Crop Management," *Science* 233:29-33 (1986).
20. Shugart, H. H. *A Theory of Forest Dynamics* (New York: Springer-Verlag, 1982).
21. O'Neill, R. V., R. H. Gardner, L. W. Barnthouse, G. W. Suter, S. G. Hildebrand, and C. W. Gehrs. "Ecosystem Risk Analysis: A New Methodology," *Environ. Toxicol. Chem.* 1:167-177 (1982).
22. O'Neill, R. V., S. M. Bartell, and R. H. Gardner. "Patterns of Toxicological Effects in Ecosystems: A Modeling Study," *Environ. Toxicol. Chem.* 2:451-461 (1983).
23. O'Neill, R. V., D. L. DeAngelis, J. B. Waide, and T. F. H. Allen. *A Hierarchical Concept of Ecosystems* (Princeton, NJ: Princeton University Press, 1986).
24. Fletcher, J. F., and W. L. Dotson. "HERMES — A Digital Computer Code for Estimating Regional Radiological Effects from the Nuclear Power Industry," Hanford Engineering Development Laboratory, Hanford, WA, HEDL-TME-71-168 (1971).
25. Haar, R., G. Swartzman, and T. Zaret. "Evaluation of Simulation Models in Power Plant Impact Assessment: A Case Study Using Lake Keowee," U.S. Nuclear Regulatory Commission, Washington, D.C., NUREG/CR-2436 (1981).
26. Shugart, H. H., and D. C. West. "Long-Term Dynamics of Forest Ecosystems," *Am. Sci.* 69:647-652 (1981).
27. Zedler, P. H., C. R. Gautier, and G. S. McMaster. "Vegetation Change in Response to Extreme Events: The Effects of a Short Interval Between Fires in California Chaparral and Coastal Shrub," *Ecology* 64:809-818 (1983).
28. O'Neill, R. V., and J. B. Waide. "Ecosystem Theory and the Unexpected: Implications for Environmental Toxicology," in *Management of Toxic Substances in Our Ecosystems*, B. W. Cornaby, Ed. (Ann Arbor, MI: Ann Arbor Science Publishers, 1981), pp. 43-73.
29. O'Neill, R. V., and J. M. Giddings. "Population Interactions and Ecosystem Function," in *Systems Analysis of Ecosystems*, G. S. Innis, and R. V. O'Neill, Eds. (Fairland, MD: International Cooperative Publishing House, 1979), pp. 103-123.
30. O'Neill, R. V., B. S. Ausmus, D. R. Jackson, R. I. Van Hook, P. Van Voris, C. Washburne, and A. P. Watson. "Monitoring Terrestrial Ecosystems by Analysis of Nutrient Export," *Water Air Soil Pollut.* 8:271-277 (1977).
31. Van Voris, P., R. V. O'Neill, W. R. Emanuel, and H. H. Shugart. "Functional Complexity and Ecosystem Stability," *Ecology* 61:1352-1360 (1980).
32. Hammons, A. S. *Methods for Ecological Toxicology: A Critical Review of Laboratory Multispecies Tests* (Ann Arbor, MI: Ann Arbor Science Publishers, 1981).
33. Giddings, J. M. "Laboratory Tests for Chemical Effects on Aquatic Population Interactions and Ecosystem Properties," in *Methods for Ecological Toxicology*, A. S. Hammons, Ed. (Springfield, VA: National Technical Information Service, ORNL-5708; EPA-560/11-80-026, 1981), pp. 23-91.
34. Suter, G. W. "Laboratory Tests for Chemical Effects on Terrestrial Population Interactions and Ecosystem Properties," in *Methods for Ecological Toxicology*, A. S. Hammons, Ed. (Springfield, VA: National Technical Information Service, ORNL-57081; EPA-560/11-80-026, 1981), pp. 93-153.
35. O'Neill, R. V., G. W. Suter, and J. M. Giddings. "Measures of Ecosystem Function," Unpublished manuscript (1981).
36. Reichle, D. E., R. V. O'Neill, and W. F. Harris. "Principles of Energy and Material Exchange in Ecosystems," in *Unifying Concepts in Ecology*, W. H. Van Dobben, and R. H. Lowe-McConnell, Eds. (The Hague, The Netherlands: W. Junk, 1975), pp. 27-43.

37. O'Neill, R. V., and D. E. Reichle. "Dimensions of Ecosystem Theory," in *Forests: Fresh Perspectives from Ecosystem Analysis*, R. H. Waring, Ed. (Corvallis, OR: Oregon State University Press, 1980), pp. 11-26.
38. Whittaker, R. H., and G. M. Woodwell. "Evolution of Natural Communities," in *Ecosystem Structure and Function*, A. Wiens, Ed. (Corvallis, OR: Oregon State University Press, 1972), pp. 137-159.
39. Odum, H. T. "Primary Production in Flowing Waters," *Limnol. Oceanogr.* 1:102-117 (1956).
40. Beyers, R. J. "Relationship Between Temperature and the Metabolism of Experimental Ecosystems," *Science* 136:980-982 (1962).
41. Beyers, R. J. "The Metabolism of Twelve Aquatic Laboratory Microecosystems," *Ecol. Monogr.* 33:281-306 (1963).
42. Copeland, B. J. "Evidence for Regulation of Community Metabolism in a Marine Ecosystem," *Ecology* 46:563-564 (1965).
43. Gorden, R. W., R. J. Beyers, E. P. Odum, and R. G. Eagon. "Studies of a Simple Laboratory Microecosystem: Bacterial Activities in a Heterotrophic Succession," *Ecology* 50:86-100 (1969).
44. Giddings, J. M., and G. K. Eddlemon. "Photosynthesis/Respiration Ratios in Aquatic Microcosms Under Arsenic Stress," *Water Air Soil Pollut.* 9:207-212 (1978).
45. Harris, W. F., B. S. Ausmus, G. K. Eddlemon, S. J. Draggon, J. M. Giddings, D. R. Jackson, R. J. Luxmoore, E. G. O'Neill, R. V. O'Neill, M. Ross-Todd, and P. Van Voris. *Microcosms as Potential Screening Tools for Evaluating Transport and Effects of Toxic Substances* (Springfield, VA: National Technical Information Service, EPA-600/3-80-042, 1980).
46. Riley, G. A. "Factors Controlling Phytoplankton Populations on Georges Bank," *J. Mar. Res.* 6:54-73 (1956).
47. Odum, H. T. "Trophic Structure and Productivity of Silver Springs, Florida," *Ecol. Monogr.* 27:55-112 (1957).
48. Odum, H. T., and C. M. Hoskin. "Comparative Studies on the Metabolism of Marine Waters," *Publ. Inst. Mar. Sci. Univ. Tex.* 5:16-46 (1958).
49. Jordan, M., and G. E. Likens. "An Organic Carbon Budget for an Oligotrophic Lake in New Hampshire, U.S.A.," *Verh. Int. Ver. Limnol.* 19:994-1003 (1975).
50. McConnell, W. J. "Productivity Relations in Carboxy Microcosms," *Limnol. Oceanogr.* 7:335-343 (1962).
51. Whitworth, W. R., and T. H. Lane. "Effects of Toxicants on Community Metabolism in Pools," *Limnol. Oceanogr.* 14:53-58 (1969).
52. Odum, H. T., and R. C. Pinkerton. "Time's Speed Regulator, the Optimum Efficiency for Maximum Power Output in Physical and Biological Systems," *Am. Sci.* 43:331-343 (1955).
53. O'Neill, R. V. "Ecosystem Persistence and Heterotrophic Regulation," *Ecology* 57:1244-1253 (1976).
54. DeAngelis, D. L. "Energy Flow, Nutrient Cycling and Ecosystem Resilience," *Ecology* 61:764-771 (1980).
55. Emanuel, W. R., H. H. Shugart, and D. C. West. "Spectral Analysis and Forest Dynamics: The Effects of Perturbation on Long-Term Dynamics," in *Time Series and Ecological Processes*, H. H. Shugart, Ed. (Philadelphia: Society for Industrial and Applied Mathematics, 1978), pp. 193-207.
56. Emanuel, W. R., D. C. West, and H. H. Shugart. "Spectral Analysis of Forest Model Time Series," *Ecol. Model.* 4:313-326 (1978).

57. Dwyer, R. L., and K. T. Perez. "An Experimental Examination of Ecosystem Linearization," *Am. Nat.* 121:305-323 (1983).
58. Dwyer, R. L., and J. N. Kremer. "Frequency Domain Sensitivity Analysis of an Estuarine Ecosystem Simulation Model," *Ecol. Model.* 18:35-54 (1983).
59. Schindler, D. W., F. A. J. Armstrong, S. K. Holmgren, and G. J. Burnskill. "Eutrophication of Lake 227, Experimental Lakes Area, Northwestern Ontario, by Addition of Phosphate and Nitrate," *J. Fish. Res. Bd. Can.* 28:1763-1782 (1971).
60. Golterman, H. L., Ed. *Interactions Between Sediments and Fresh Water* (Wageningen, The Netherlands: Dr. W. Junk, 1976).
61. Hutchinson, G. E. *A Treatise on Limnology, Volume 1* (New York: John Wiley & Sons, 1975).
62. Mortimer, C. H. "The Exchange of Dissolved Substances Between Mud and Water in Lakes," *J. Ecol.* 29:280-329 (1941).
63. Mortimer, C. H. "The Exchange of Dissolved Substances in Lakes," *J. Ecol.* 30:147-201 (1942).
64. Pomeroy, L. R., E. E. Smith, and C. M. Grant. "The Exchange of Phosphate Between Estuarine Water and Sediment," *Limnol. Oceanogr.* 10:167-172 (1965).
65. Porcella, D. R., V. D. Adams, and P. A. Cowan. "Sediment-Water Microcosms for Assessment of Nutrient Interactions in Aquatic Ecosystems," in *Biostimulation and Nutrient Assessment*, E. Middlebrooks, D. H. Falenborg, and T. E. Maloney, Eds. (Ann Arbor, MI: Ann Arbor Science Publishers, 1976), pp. 293-322.
66. Pritchard, P. H., A. W. Bourquin, H. L. Frederickson, and T. Maziarz. "System Design Factors Affecting Environmental Fate Studies in Microcosms," in *Microbial Degradation of Pollutants in Marine Environments*, A. W. Bourquin and P. H. Pritchard, Eds. (Springfield, VA: National Technical Information Service, EPA-600/9-79-012, 1979).
67. Domsch, K. H. "Effects of Fungicides on Microbial Populations in Soil," in *Pesticides in the Soil* (East Lansing, MI: Michigan State University Press, 1970), pp. 42-46.
68. Coughtrey, P. J., C. H. Jones, M. H. Martin, and S. W. Shales. "Litter Accumulation in Woodlands Contaminated by Pb, Zn, Cd, and Cu," *Oecologia* 39:51-60 (1979).
69. Jackson, D. R., and A. P. Watson. "Disruption of Nutrient Pools and Transport of Heavy Metals in a Forested Watershed Near a Lead Smelter," *J. Environ. Qual.* 6:331-338 (1977).
70. Lighthart, A., G. H. Bond, and M. Richard. "Trace Element Research Using Coniferous Forest Soil/Litter Microcosms," (Corvallis, OR: U.S. Environmental Protection Agency, EPA-6000/3-77-091, 1977).
71. Ruhling, A., and G. Tyler. "Heavy Metals Pollution and Decomposition of Spruce Needle Litter," *Oikos* 24:402-416 (1973).
72. Spalding, B. P. "The Effect of Biocidal Treatments on Respiration and Enzymatic Activities of Douglas-Fir Needle Litter," *Soil Biol. Biochem.* 10:537-543 (1978).
73. Stotzsky, G. "Microbial Respiration," in *Methods of Soil Analysis, Part 2, Chemical and Microbial Properties*, C. A. Black, Ed. (Madison, WI: American Society of Agronomy, 1965), pp. 1550-1572.
74. Tyler, G. "Heavy Metal Pollution, Phosphate Activity, and Mineralization of Organic Phosphorus in Forest Soils," *Soil Biol. Biochem.* 8:327-332 (1976).
75. Baas Becking, L. G. M., I. R. Kaplan, and D. Moore. "Limits of the Natural Environment in Terms of pH and Oxidation-Reduction Potential," *J. Geol.* 68:243-284 (1960).

76. Stumm, W. and J. J. Morgan. *Aquatic Chemistry* (New York: John Wiley & Sons, 1970).
77. Schindler, J. E., J. B. Waide, M. C. Waldron, J. J. Haines, S. P. Schreiner, M. L. Freedman, S. L. Benz, D. P. Pettigrew, L. A. Schissel, and P. J. Clarke. "A Microcosm Approach to the Study of Biogeochemical Systems. 1. Theoretical Rationale," in *Microcosms in Ecological Research*, J. P. Giesy, Ed. (Aiken, SC: Savannah River Ecology Laboratory, 1980).
78. Waide, J. B., J. E. Schindler, M. C. Waldron, J. J. Haines, S. P. Schreiner, M. L. Freedman, S. L. Benz, D. P. Pettigrew, L. A. Schissel, and P. J. Clarke. "A Microcosm Approach to the Study of Biogeochemical Systems. 2. Responses of Aquatic Laboratory Microcosms to Physical, Chemical, and Biological Perturbations," in *Microcosms in Ecological Research*, J. P. Geisy, Ed. (Aiken, SC: Savannah River Ecology Laboratory, 1980), pp. 204-223.
79. Webster, J. R. "Analysis of Potassium and Calcium Dynamics in Stream Ecosystems on Three Southern Appalachian Watersheds of Contrasting Vegetation," PhD Thesis, University of Georgia, Athens (1975).
80. Webster, J. R., and B. C. Patten. "Effects of Watershed Perturbation on Stream Potassium and Calcium Dynamics," *Ecol. Monogr.* 19:51-72 (1979).
81. Wallace, J. B., J. R. Webster, and W. R. Woodall. "The Role of Filter Feeders in Flowing Waters," *Arch. Hydrobiol.* 79:506-532 (1977).
82. Elwood, J. W., J. D. Newbold, R. V. O'Neill, R. W. Stark, and P. T. Singley. "The Role of Microbes Associated with Organic and Inorganic Substrates in Phosphorus Spiralling in a Woodland Stream," *Verh. Int. Ver. Limnol.* 21:850-863 (1981).
83. Elwood, J. W., J. D. Newbold, R. V. O'Neill, and W. Van Winkle. "Resource Spiralling: An Operational Paradigm for Analyzing Lotic Ecosystems," in *Dynamics of Lotic Ecosystems*, T. D. Fontaine, and S. M. Bartell, Eds. (Aiken, SC: Savannah River Ecology Laboratory, 1983), pp. 3-27.
84. Newbold, J. D., J. W. Elwood, R. V. O'Neill, and W. Van Winkle. "Measuring Nutrient Spiralling in Streams," *Can. J. Fish. Aquat. Sci.* 38:860-863 (1981).
85. Newbold, J. D., P. J. Mulholland, J. W. Elwood, and R. V. O'Neill. "Organic Carbon Spiralling in Stream Ecosystems," *Oikos* 38:266-272 (1982).
86. Newbold, J. D., R. V. O'Neill, J. W. Elwood, and W. Van Winkle. "Nutrient Spiralling in Streams," *Am. Nat.* 120:628-652 (1982).
87. Newbold, J. D., J. W. Elwood, R. V. O'Neill, and A. L. Sheldon. "Phosphorus Dynamics in a Woodland Stream Ecosystem: A Study of Nutrient Spiralling," *Ecology* 64:1249-1265 (1983).
88. Odum, E. P. "The Strategy of Ecosystem Development," *Science* 164:262-270 (1969).
89. Likens, G. E., F. H. Bormann, R. S. Pierce, J. S. Eaton, and N. M. Johnson. *Biogeochemistry of a Forested Ecosystem* (New York: Springer-Verlag, 1977).
90. Jackson, D. R., J. J. Selvidge, and B. S. Ausmus. "Behavior of Heavy Metals in Forest Microcosms. Transport and Distribution Among Components," *Water Air Soil Pollut.* 10:3-11 (1978).
91. Jackson, D. R., B. S. Ausmus, and M. Levine. "Effects of Arsenic on Nutrient Dynamics of Grassland Microcosms and Field Plots," *Water Air Soil Pollut.* 11:13-21 (1979).
92. Jackson, D. R., and J. M. Hall. "Extraction of Nutrients from Intact Soil Cores to Assess the Impact of Chemical Toxicants on Soil," *Pedobiologica* 13:272-278 (1978).

CHAPTER 5

Decision Analysis as a Tool in Integrated Environmental Management

Randall A. Kramer

INTRODUCTION

Recently, the town of Whately, MA, was faced with an increasingly common dilemma. Water monitoring revealed that an aquifer used by some of its residents for drinking water was contaminated by pesticides. The pesticides posed health risks of increased probabilities of certain types of cancers to a small number of households. When economists at the University of Massachusetts calculated these health risks monetarily using conventional methods, they estimated that continued use of the aquifer would result in losses over time of about $50,000. The cost of developing a new water supply for the 200 affected residents was $4 million.[1] Despite this unfavorable balance of benefits and costs, the citizens and their elected officials chose to proceed with the development of a new water supply.

Why would the town favor a policy decision with such an unfavorable benefit-cost ratio? People react to risk in ways that are difficult to model and predict. In this particular case, the citizens may have misperceived the degree of risk involved, i.e., they may have thought that the probability of health effects was much greater than scientific evidence suggested. Alternatively, they may have been acting in a highly risk averse manner, i.e., they may have realized that the health risks had very low probabilities, but because the roll of the dice meant that death might come sooner to one of their own family members, they were willing to pay a great deal to avoid something with a very low likelihood of occurrence.

This example illustrates some of the difficulties that arise when environmental management decisions are subject to uncertainty. In environmental policy, uncertainty is receiving considerable attention due to such issues as nuclear plant safety, the handling and disposal of hazardous waste, and the control of toxic substances in food supplies.

Government agencies are increasingly faced with risk assessment and risk management questions, and society must decide how much it is willing to spend to reduce risks. Reports from Bhopal, India of lax safety procedures suggest that Union Carbide was not spending enough to prevent accidents at its chemical plant. But how far should risks be reduced? Union Carbide probably could have done more to reduce the risk of a major release of toxic chemicals, but it is doubtful that the industry could have reduced to zero the probability of an accident. How safe is safe enough, and what is it worth?

Decision analysis has a role in integrated environmental management (IEM). It is an interdisciplinary field that draws from psychology, economics, statistics, operations research, and other disciplines to either explain or aid decision making.[2] Decision analysis has prescriptive or normative uses, i.e., it can play a role in helping environmental decision makers make more informed decisions. This discussion reviews basic concepts of decision analysis and focuses on two types of models: expected utility and safety first.

THE IMPORTANCE OF RISK AND UNCERTAINTY IN ENVIRONMENTAL MANAGEMENT

In a book published in 1921,[3] the economist Frank Knight made a now-famous distinction between risk and uncertainty: risk is what can be quantified, while uncertainty is what cannot be quantified. If this distinction is followed, then anything for which a probability distribution can be developed to indicate the likelihood of occurrence should be called risk.

A few decades later, however, the statistician L. J. Savage argued that what was important in decision making was the concept of subjective probability.[4] Subjective probability is the degree of belief that an individual attaches to a particular outcome. Savage argued that people make probabilistic judgments all the time with limited or nonexistent historical or other so-called objective probability data. As a result, the earlier distinction between risk and uncertainty was blurred, and much of the literature over the past 2 decades has used the two terms interchangeably.

However, in recent years, the distinction between risk and uncertainty has made somewhat of a comeback because of the perplexing technological risks, many of them environmentally related, that are emerging in society. Some of these risks are so complex and lay people or scientists have so little experience with them that neither are able to assign them subjective probabilities. This discussion considers only the sources of randomness that can be quantified, for pragmatic reasons. In formulating quantitative environmental management models, only quantifiable risks can be factored into the models. The sources of nonquantifiable uncertainty should be recognized, but cannot be treated directly with decision analysis. The terms "risk" and "uncertainty" are used interchangeably here.

Research on the management of environmental resources is increasingly concerned with risk.[5] Why has risk received so much attention in recent years? In part, this reflects a shift in the way society views decision making. There is less of a tendency to view the world as deterministic. In the environmental area, health researchers are finding that the main effect of some types of pollution is to increase the probability that individuals will contract a debilitating disease during their lifetime. These environmental risks are increasingly being communicated to the public, and as the public becomes more adept at viewing the world as stochastic, there is growing demand for public and private environmental managers to incorporate risk in their decision making. Managers are turning to the scientific community for help in dealing with these issues.

Risk can affect environmental decision making for a number of reasons. Some sources of risk are nature related. First, for some renewable resources, resource productivity relies on a biological production process that is naturally stochastic. For example, timber production can be affected by random disturbances due to pests and fire, and fish and wildlife populations fluctuate due to weather and disease. Reportedly, the dolphin population along the Atlantic Coast declined significantly last year due to the effects of red tide. Second, environmental pollution has a stochastic component due to weather. Typically, nonpoint source water pollution increases dramatically during rainy months of the year. Similarly, air quality diminishes dramatically during periods of temperature inversion.

Other sources of risk affecting environmental decision making arise from human activity. Both pollution and resource extraction fluctuate in response to economic cycles. Furthermore, governmental policies that increase or reduce economic risks can also influence natural resource use. For example, it is often argued that federally subsidized flood insurance has encouraged coastal development by reducing the risk of real estate investments. Development has brought increased pressure on coastal ecosystems. Finally, long-term environmental decision making is complicated by the risk of future resource demand. At the current time, the public may not be interested in conserving economically unimportant plant or animal species. However, the demand in 10 or 20 years for protecting the species may be greater, either because of the discovery of new economic uses or because of a shift in values that leads to a greater appreciation of biodiversity. Due to the irreversibility of many types of environmental decisions, future changes in resources demand are critically important to consider.

DECISION TREES

The outcome of environmental decision making in both the public and private sector depends not only on the choices made but also on external events that are beyond the control of the decision maker. This makes decision making more difficult, since each option under consideration can have a variety of outcomes. A decision tree can assist decision makers in structuring the decision making process[2] through consideration of the various choices, events, and outcomes that are possible. A decision tree specifies the interactions between the decision maker and the environment and provides a compact representation of the range of scenarios. In order for a decision tree to be useful, it generally must simplify reality. Only the most important choices and possible events should be considered so as to keep the decision tree manageable. Applications of decision trees to environmental management issues include studies of rehabilitation of damaged ecosystems[6] and management of endangered species.[7]

A decision tree can be illustrated by considering a simple, hypothetical problem representing a decision faced by a coastal management agency. The agency must decide on issuing a permit to allow development in a pristine coastal area. If development is allowed to proceed, it will spur the local economy, leading to new employment opportunities, greater earnings for local businesses, and higher tax revenues for local governments. However, developing the area without adequate pollution control could lead to impaired water quality and closed shellfish beds. This will lead to income losses for local watermen and will depress the local economy. If the permit is not issued, the status quo will be maintained in terms of the economy and the water quality. Thus, the primary gains will be new economic activity, and the primary losses will be damage to the shellfish industry.

The decision problem facing the agency is described with the decision tree in Figure 1. The decision involves two possible actions: issue or deny a permit for development. If the permit is issued, the first branch of the tree leads to the external event of adequate or inadequate pollution control. If pollution is properly managed, water quality is maintained. The alternative event is that despite whatever conditions are attached to the permit, pollution control may turn out to be inadequate, e.g., septic systems may fail.

Depending on which event occurs, the action chosen by the agency will have very different outcomes. If pollution control is adequate, the gains will be the $500,000 annually in new economic activity (after accounting for private and government costs of generating that activity). There will be no losses to the shellfish industry, so the net gains will also be $500,000. Alternatively, if factors beyond the control of the agency cause water quality to decline, the new economic activity will generate $500,000 annually, but shellfish beds will be closed and the associated economic losses will be $1.5 million annually. The net outcome of inadequate pollution control will be losses of $1 million.

Up to this point, nothing has been said about the chances of the two events occurring. Clearly the consequences of the agency's decision hinge on those probabilities. Suppose that by examining past permit decisions and subsequent outcomes it is possible to assign a probability of 0.8 to the event that adequate pollution control will be achieved. With this additional information, the expected value of issuing the permit can be computed. The expected value is the weighted average of the two outcomes, using the probabilities as weights:

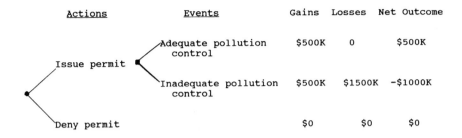

Figure 1. The decision tree.

EV (permit) = (0.8 x $500,000) + (0.2 x -$1,000,000) = $200,000.

Comparing the $200,000 expected value of issuing the permit to the alternative of denying the permit, which has an expected value of zero, the agency would learn from this analysis that the expected gains from the permit are positive. This simple illustration shows that a decision tree can be a useful decision aid.

Clearly, the tree could be more complex. The number of events could be increased to reflect different levels of degradation. Also, the tree could be expanded into multiple stages to allow later actions by the agency to require cleanup if pollution control turns out to be inadequate. Yet, various elaborations of the tree would not address one important shortcoming: the agency may not want to base its decisions on expected outcomes alone. For example, the board may decide to deny the permit despite the expected $200,000 gains because it does not feel that these modest gains are sufficient to offset the risk of financial ruin to the shellfish industry. Models can allow decision makers to consider tradeoffs between higher expected gains and higher levels of risk.

MODELS

The analyst who wishes to go beyond a decision tree to develop a more elaborate decision aid can choose from several frameworks. Two frameworks are suitable for environmental decision making: expected utility analysis and safety-first analysis.

The expected utility model is one of the most widely used models of environmental decision making under risk. The expected utility model has a long history.[8] Early 18th century mathematicians were puzzled that in certain real and hypothesized situations people did not seem to follow the decision rule of maximizing expected payoff. One of these mathematicians, Daniel Bernoulli, proposed that when people are faced with risky choices, they do not act to maximize expected payoffs, but rather seek to maximize expected utility (although he used the term "moral wealth" rather than expected utility). Bernoulli's principle, which subsequently became the expected utility theorem, was later proven by von Neumann and Morgenstern[9] in their 1944 book *Theory of Games and Economic Behavior*.

In its modern form, the expected utility model assumes that individuals choose between risky alternatives based on their beliefs and preferences. The beliefs relevant in this context are the perceived chances of the occurrence of the uncertain events that impact the consequences of the actions. These beliefs are encoded mentally as subjective probabilities. Preferences in the expected utility model are the relative rankings that individuals assign to different consequences. These preferences can be represented by appropriately elicited utility functions.

According to the expected utility model, a rational decision maker will choose the action whose subjective probability distribution of consequences maximizes the decision maker's expected utility. More pragmatically, this means that decision makers weigh expected returns and risks to select alternatives that will give them the highest level of satisfaction. In the context of environmental risk, the expected utility model implies that risk-averse individuals will be willing to pay, through higher taxes or higher prices for goods, to reduce environmental hazards. There is a limit to what they will be willing to pay. Therefore, through their collective actions, their society will not attempt to completely eliminate environmental risks, even if it was technically possible to do so.

Psychologists have challenged the rational view of decision making that the expected utility model represents.[10] They argue that many people are not capable or willing to assign subjective probabilities to a wide range of alternatives and then systematically evaluate the alternatives according to a well-defined preference structure. Instead, most decision making under risk relies on heuristics.

One decision aid that relies on heuristics and has been widely used is the safety-first model. Unlike the expected utility model, it is not derived from a set of axioms, yet it has considerable intuitive appeal and is consistent with the concept of a hierarchal goal structure.[11] The safety-first model postulates that individuals first seek to satisfy a preference for safety and then seek to satisfy some other, usually monetary, goal.

In the context of environmental risk, a safety-first model implies that individuals will place the highest priority on avoiding environmental risk, but once the risk is contained at a satisfactory level, they will seek to achieve other goals as well. For example, because of the uncertainty and irreversibility associated with managing endangered species, a public decision maker might adopt a safe minimum standard approach to project selection.[12] In considering the alternative development projects, the decision maker would avoid a project that raised the probability of a species moving into the critical zone where further depletion is irreversible. This rule would be followed unless the social costs of it were unacceptably large.

MANAGEMENT OF GROUNDWATER QUALITY

An example of decision analysis used in environmental decision making is some recent research on groundwater management conducted at Virginia Tech. Ground water serves as a drinking water source for most of the rural population in the United States. Mounting evidence indicates that the primary economic activity in rural areas, agriculture, is contributing to the degradation of groundwater supplies in some areas of most states. Groundwater pollution in rural areas is primarily from pesticides and nitrates.

A study was undertaken by a group of economists and engineers to investigate alternative public policies for implementing controls on nitrate pollution from agricultural sources.[13] Rockingham County, Virginia, was selected as a case study because of reports of excessive groundwater nitrate levels attributed to agricultural land uses. Much of the excessive infiltration of nitrates was believed to originate from dairy farms, so the analysis focused on the county's dairy industry.

A mathematical programming model was constructed that linked together economic decision making with a water quality model to simulate the effects of changes in public policies. The model included simulated nitrate loadings under a variety of policy scenarios and land management practices. These nitrate loadings could vary greatly from year to year due to rainfall variation, manure spreading timetables, and variability in management practices. Initial analysis indicated that the nitrate loading factors could vary up to 95% of their mean values. This suggested that any groundwater management program that established management strategies based on average nitrate loadings could be in violation of the intended standard a high percentage of the time. As a result, pollution uncertainty was incorporated directly into the model.

The analysis made use of a safety-first approach with a mathematical programming technique known as chance-constrained programming. The model first ensured that the safety constraint, expressed in terms of improved water quality, was met a high percentage of the time, and then proceeded to select land use practices that would be the most profitable under this constraint. From a decision-analysis framework, the safety constraint can be viewed as an external regulation imposed on the farmer by a risk averse regulator. Alternatively, the safety-first constraint can be interpreted as a self-imposed risk limit if the farmer is operating under a strict liability rule for contamination, as was recently legislated in Connecticut.

To illustrate the effects of uncertainty on environment management, a few of the results of this study are shown in Table 1. One environmental goal examined with the model was a 40% reduction in nitrate infiltration. This number corresponded with a 40% reduction in nitrogen-related pollution of the Chesapeake Bay sought by the governors of Virginia, Maryland, and Pennsylvania. The first column shows that requiring the representative dairy farm to reduce nitrate leaching by 40% leads to a modest income decline of 1.6%. However, this assumes that nitrate leachings are nonstochastic. A reasonable certainty of achieving the reduction imposes much higher economic costs. If an 80% safety level is attached to the management strategy, economic costs (as measured by income declines) increase 4.9% from the base case. A 90% safety level increases economic costs per farm by 13%.

CONCLUSIONS

Government agencies and private citizens are increasingly faced with risk assessment and risk management questions. The questions generally relate to the

Table 1. The Effects of Uncertainty on Achieving a 40% Reduction in Nitrate Contamination of Ground Water by a Representative Dairy Farm

	Base Case	40% Pollution Reduction		
		Uncertainty Ignored	80% Safety Level	90% Safety Level
Average income	$55,407	$54,500	$52,714	$48,397
Percent decline		1.6	5.9	13
Average NO$_3$ leached	2682	1623	1021	750
Percent decline		40	62	72

degree of safety desired for the environment and the amount society is willing to pay for safety. Given the pervasiveness of uncertainty affecting environmental resources, it is important that decision makers recognize the stochastic nature of their decision context. Decision analysis can assist in formulating a structured approach to decision making that formally introduces sources and degrees of risk and effects of actions to mitigate risk.

REFERENCES

1. Harper, C. R., and C. E. Willis. "Economic Advice and Public Decisions," *Choices* First Quarter, p. 33 (1989).
2. von Winterfeldt, D., and W. Edwards. *Decision Analysis and Behavioral Research* (Cambridge, MA: Cambridge University Press, 1986).
3. Knight, F.. *Risk, Uncertainty and Profit* (London: Houghton-Mifflin, 1921).
4. Savage, L. J. *The Foundations of Statistics* (New York: John Wiley & Sons, 1954).
5. Batie, S. S., and R. A. Kramer, Eds. *Risk and Natural Resources*, Proceedings of a Workshop. SNREC Publication No. 24 (Mississippi State: Mississippi State University Southern Rural Development Center, 1987).
6. Maguire, L. A. "Decision Analysis: An Integrated Approach to Ecosystem Exploitation and Rehabilitation," in *Rehabilitating Damaged Ecosystems, Vol. II*, J. Cairns, Jr., Ed. (Boca Raton, FL: CRC Press, 1988), pp. 105-122.
7. Maguire, L. A. "Using Decision Analysis to Management Endangered Species," *J. Environ. Manage.* 22:345-360 (1986).
8. Lifson, M. W. *Decision and Risk Analysis for Practicing Engineers* (Boston: Barnes & Noble, 1972), pp. 57-59.
9. von Neumann, J., and O. Morgenstern. *Theory of Games and Economic Behavior* (Princeton, NJ: Princeton University Press, 1947).
10. Hogarth, R. *Judgement and Choice: The Psychology of Decisions* (New York: John Wiley & Sons, 1980).
11. Kunreuther, H. "Economic Analysis of Natural Hazards: An Ordered Choice Approach," in *Natural Hazard Perception and Choice*, G. F. White, Ed. (London: Oxford University Press, 1974).

12. Bishop, R. C. "Endangered Species and Uncertainty: The Economics of a Safe Minimum Standard," *Am. J. Agric. Econ.* February:10-18 (1978).
13. Halstead, J. M., S. S. Batie, D. B. Taylor, C. D. Heatwole, P. L. Diebel, and R. A. Kramer. "Managing Agricultural Nitrate Contamination of Ground Water: Impacts of Uncertainty on Policy Formulation," paper presented at the American Agricultural Economics Association Annual Meeting, Baton Rouge, LA, July 31-August 2, 1989.

CHAPTER 6

Applied Ecology, Its Practice and Philosophy

L. B. Slobodkin and D. E. Dykhuizen

INTRODUCTION

Keeping the environment pristine is impossible due to human activities that change ecosystems — deforestation, eutrophication, and desertification. Did the inhabitants of ancient Babylon or Petra or the cities of the Mayans so degrade their environment that the social order collapsed and the cities died? Will chemical pollution force civilization to abandon large tracts of land to small rural groups because the environment can no longer support cities? This nightmare is so haunting that some sophisticated citizens routinely attempt to block all development in their neighborhoods. These people are sufficiently numerous to merit an acronym: NIMBYS ("Not In My Back Yard").

Clear thought is needed, but some ways of thinking impede answers. The perception of the environment and its inhabitants must be clarified. This chapter focuses on four concepts that are not mutually exclusive: a medical analogy; a discussion of the differences among scientists, technologists, and practitioners; how to gain information to exercise control of the environment; and finally, an example of a specific set of problems in ecotoxicology.

THE MEDICAL ANALOGY

An implicit analogy with medicine emerges from consideration of the health of an ecosystem. Sufficiently large quantities of a variety of chemical pollutants can "kill" an ecosystem, including its human inhabitants. In addition, chemical concentrations that are neither obviously disastrous nor clearly innocuous can affect the "health" of an ecosystem.

Medical symptoms depend in part on the patients' perceptions. Within living memory, obesity and alcoholism were issues of morality, but now they are considered medical problems. As in medicine, so in applied ecology, the subject matter of the field arises from concern with the urgency of problems, not necessarily from an ability to solve them. Just as what physical health means to the individual (freedom from pain, low chance of dying, good physical shape) can vary, definitions of ecosystem health also depend in part on desires and perceptions. The job of applied ecology, like medicine, is to answer to the expressed needs of people and simultaneously pinpoint needs and dangers that might not be visible to the layman.

Unfortunately, as in medicine, there is a requirement to do something even in the absence of appropriate technology and knowledge. This was dramatically illustrated during the 1989 Alaskan oil spill. The best will in the world, as well as a fair share of the money and an abundance of resources, was available, but there was very limited technical understanding of what to do once the oil had spilled. It resembled, in this sense, the initial response to the AIDS problem.

How can the declining health of an ecosystem be determined? Will the species diversity decrease, the primary productivity increase or decrease, the stability decrease? Unfortunately, no agreement has yet surfaced on the general value of any particular index[1] since it is not simple to choose criteria of ecological health. Are swamps unhealthy ecosystems because they are unhealthy to humans? Using people as one of the ecotoxicological indicator species is not unreasonable since good statistics are available on this species, but using human health as the sole criteria for ecosystem health would be unfortunate.

Is the distinction between acute and chronic illness useful for applied ecology? Spills of toxic chemicals or oil, as in Alaska, are acute problems. The proper response must be devised at the same time as the best available response is being implemented. On the other hand, chronic releases of chemicals can, over an extended period of time, destroy an ecosystem, as acid rain seems to be destroying the Black Forest.

Even in medicine, health is a nebulous concept compared to illness. A person in generally poor physical condition may be defined as healthy because nothing is specifically wrong, while a person of high athletic ability and excellent musculature, but with recognizable arteriosclerosis or high blood pressure, is not healthy because probable causes of death can be identified. Who is healthier — the athlete on steroids, the jogger with knee trouble, or the man who cannot catch his breath after running for a bus? A person with an elevated temperature is usually defined as ill because of the general association between elevated temperature and many kinds of infections, but lack of elevated temperature does not imply health. Medicine does not require complete correlations; partial correlations are useful. The same should be true for ecology.

Health cannot be quantitated in either medicine or ecology. A 19th century observer would have decided that the gigantic populations of the now-extinct passenger pigeon seemed ecologically healthier than those of the persistent Indian lion.

Most ecosystems free of anthropogenic effects will not change very quickly, but they do change slowly from one recognizable state to another. When has an ecosystem died? Is ecosystem death merely a change to another ecosystem, as when a rain forest is converted into a coarse grass prairie by human interference or a microbial ecosystem in a sewage treatment plant changes over? Therefore, the ultimate test of lack of health, namely death, cannot be simply transferred from medicine to ecology.

"Health" (or some similar term) of an ecosystem is a real concept, but this health is not easily measured. Medical terms can be used to express concern, but they should not be used as quantitative terms for scientific discussion, writing of legislation, or policy decisions.

Ecology and medicine are alike as both are confronted with the problems of a subject matter area rather than being free to choose problems for their intrinsic interest, solvability, or intellectual charm.[2] In like fashion, ecology ought to share with medicine the ability to learn from the conditions it is treating.

SCIENCE, TECHNOLOGY, AND PRACTICE

If the question of what effect is desired for a particular ecosystem is clear, then deep philosophical and political issues about the interactions between the traditional roles of science, practice (engineering), and technology must be addressed before an effective solution can be found to many problems of applied ecology.

Science is concerned with creating an intellectual model of the material world. Technology is concerned with procedures and tools and their general use to gain or use knowledge. Practice is concerned with how to treat individual cases. Confusing the three can be dangerous.

In the realm of medicine, the scientist is concerned with such questions as how does the immune system work and how do bacterial pathogens bypass this system, what is cancer, etc. The practitioner, who in the physical sciences is called an engineer and in medicine is often symbolized by the family doctor, has to solve the problems related to a particular case, helping a particular person regain (or keep) health. He has to diagnose the problem and decide what is the best treatment based on present knowledge. Between the scientist and the practitioner is the technologist, sometimes called an engineer and sometimes called a scientist. The medical technologist develops and evaluates techniques ranging from new forceps to new drugs to NMR machines.

The three main approaches must also be considered in applied ecology. Many of the practitioners in applied ecology have almost no training in academic ecology. They are well trained in physics and chemistry in which the intellectual focus is determined by theoretical tractability. These practitioners are most comfortable with a physical sciences model, often linear, simplified, and free of feedback mechanisms, i.e., deeply nonbiological. There seems to be little regular feedback from ecological engineering to biological scientists; in addition, practitioners are not required to follow developments in more academic sciences.

Ecology is a relatively new science which, until recently, was thought of as having no technological implications. The lack of daily immersion in the urgency of applied problems which characterizes most academic ecologists may permit a broader viewpoint and framework for discussion, but the insights of the scientist, the environmentalist, the bureaucrat, and the manager are all required if actual solutions are to be generated. If practitioners in applied ecology feel that academic ecologists have no understanding of the real world as they know it, then the complexity of particular examples or case studies are not becoming part of the basic shared knowledge of the field. On the other side, practitioners are often using models of the world created by now-dead ecologists. The creative interaction between scientists, technicians, and practitioners, which actually occurs in medicine, is not yet present in applied ecology.

Scientists need to know what is happening at the applied level. Likewise, engineers need to integrate a continuing understanding of the full complexity of the natural world into their practices. This communication will not grow naturally; consider some clarifying examples. People working in fermentation and sewage treatment still use pre-Darwinian terms for their phenomena and ecologists are usually not interested in these phenomena despite their obvious relevance to ecological theory. While communication between conservation biology and academic ecology is better, university ecologists do not seem to be aware of, or influenced by, the literature surrounding environmental impact statements or questions of environmental law.

The situation is complicated by the fact that regulatory agencies, powerful environmental miscreants, and scientists concerned with toxicology and environmental impacts are entangled in a web of financial, political, philosophical, and legal antagonisms and dependencies. Are a scientist whose funding source is industry, an environmentalist whose career depends on controversy, a bureaucrat who is punished only for making a mistake, or a factory manager who must make the next paycheck capable of seeing through the fog of issues?

This introduction has attempted to view the organization of the field of applied ecology and envision how to make it productive. Integration of the area of concern requires a better defined vision than provided here, better ideas of what needs to be done, and the political will of leading practitioners of the various fragments to start the process. However, certain proposals will promote integration.

THE IMPORTANCE OF DISASTERS

If an investigator has enough knowledge of an ecosystem to recognize when deleterious changes occur, then damage to a particular ecosystem can be identified.

However, people claiming a deep understanding of a particular ecosystem usually have neither the data nor the credentials to influence litigation.

As discussed above, there are no extrinsic criteria by which damage to a particular ecosystem can be determined unless the change is so obvious and fast that the average disinterested person can see it. If the damage is less obvious, how can it be determined? Identical ecosystems for controlled comparison could assist in this determination. However, criteria have not been agreed upon of when similarity between ecosystems is sufficient for one of them to serve as a control, except perhaps for some microbial systems (e.g., the rumens of cows or sewage treatment plants).

If sufficient baseline data were available from a single ecosystem, the properties of the system before and after the disturbance could perhaps be compared. Then the observed changes could be evaluated for deciding what constituted a deterioration. If deterioration had, in fact, occurred then the question of torts or legal restriction or rehabilitation would arise. The practical difficulties are obvious. Baseline data of sufficient quality are generally not available and requiring that such data be available for the system that is about to be disturbed is unrealistic. Also, the meaning of deterioration often hinges on viewpoint.[2]

Ecosystems do not exist in large numbers, so an epidemiologic (i.e., statistical) approach (which is common in medicine) is not possible. It is necessary to take advantage of disasters to advance knowledge by providing internal controls for emergency treatments.

For example, assume that there will be another oil spill, regardless of how distasteful that seems. If the area affected by the Exxon Valdez oil spill had been subdivided into different treatment areas, including a control area where no attempt at cleaning had been made, and if each of these areas were monitored carefully for the next several years, researchers might have been in a position to improve the understanding of what oil spills actually do, how different treatments of oil spills affect ecosystems, and how best to prepare for the next oil spill. The AIDS epidemic initiated research programs that should eventually control the problem. Unfortunately, the next major oil spill will be met with the same lack of knowledge as the last one. A lesson from medicine that must be applied to ecology is the organization of research effort, particularly in confrontations with disasters.

Disasters must deliberately be used as experimental opportunities to test different response procedures, even in areas where nothing is done as a control. During the testing of a new medicine, some ill people are treated and some are not in determining the effectiveness of the treatment. This procedure is accepted (with some reluctance for some of the treatments for AIDS). It is quite possible that what is done is worse than doing nothing — a possibility not congenial to the entrepreneurial spirit. Not doing something everywhere will be politically unpopular, but the public should be educated on the need for controls.

Obviously, waiting for disasters is not sufficient. Model systems that allow making tentative plans for dealing with, and learning from, disasters must be studied in advance. This requires a theory of how model systems relate to natural ecosystems and how to extrapolate from one ecosystem to another.

The Superfund is "throwing money" at the same problems, but little is being learned from it. As aging research has shown, allocating money in small bundles in an area of concern can rapidly promote much improved research and techniques. More concern for society and its future should be shown by many of the leading organizations, not just the oil companies. Sometimes society does not know what it is doing, but is following such procedures so that next time the outcome will be better. Just as in medicine, the free market is not the place for basic experimentation, even if the free market is most efficient at distributing the best techniques once they are known.

AMBIGUITY OF ECOTOXICOLOGY

One illustration of a problem in applied ecology shows how appealing, and apparently reasonable, approaches must be cautiously assessed, lest they lead to endless litigation rather than to control of environmental problems. In practice, ecotoxicology can be used in three general ways to assess ecological damage,[3] apart from the use of ecotoxicology simply to monitor release of chemical pollutants. Each approach requires a different, superficially plausible assumption.

1. The most sensitive species may serve as a surrogate for the community of species. This assumes that if the most sensitive species in an ecosystem is not disturbed by a chemical change, none of the others will be. This approach does guarantee safety, if the most sensitive species can be identified. In practice, even if the most sensitive species can be tested in such a way as to mimic its field responses (a very hard response to demonstrate), it may well be argued that loss or damage to this most sensitive species need not indicate "serious" damage to the whole ecosystem and that prohibiting or restricting a particular activity on the basis of the most sensitive species response may be an excessive impediment to economic development. However, if the most sensitive species is not damaged, this is evidence that no damage has been done. There is, therefore, an oddly asymmetrical pattern to how the most sensitive species analysis may be used in legal actions, since it can be used to prove no damage but it is very difficult to prove that significant damage has occurred. It is to be expected that potential polluters (manufacturers, power plant and sewage works operators, etc.) should strongly favor most sensitive species analysis, but that conservation agencies should not.
2. The effect of a toxicant on valuable species may be used to assess damage. This assumes that in any ecosystem there are certain clearly valuable species to be preserved and other nonvaluable species that can be lost. For example, if there is popular concern for striped bass, then the effect on striped bass may be tested. This would be a valid approach except for the above-mentioned difficulty of mimicking field conditions in a test situation and for the clear fact that no species lives in isolation. A test on individuals of a particular species gives almost no information about the fate of the population of that species in nature unless the effect of the same toxicant is known for those species important to it, as either food organisms, competitors, or predators. Which species are important to which other species is often not really clear, and once again, the opportunity for litigation arises.

3. Some assume that since no one species is an adequate indicator of damage, a "microcosm" consisting of a multispecies subsample of the ecological system will provide better information. This approach attempts to retain the interactive properties of the system while making them tractable to study. Unfortunately, the resultant information remains legally equivocal. If a particular microcosm has or has not been apparently damaged by some particular treatment or set of conditions, it remains possible to argue that this is not truly representative of events in nature. A microcosm may be used as a surrogate for a natural community as a matter of convenience, but the similarity to nature may not be strong enough to permit confident extrapolation.

While the ecotoxicological approach to assessment of ecological damage may provide scientifically interesting insights, its use is limited to special cases where the general ecology of the system is well known.

Ecotoxicology can be considered in an alternative way. Organisms can act as surrogate chemists to determine particular chemical concentrations. In this case, the organisms used do not have to be part of the ecosystem under consideration. Consider the classic example of bringing canaries into coal mines in the 19th century. If the birds showed distress or died, then there was danger and further action was required. There is no role for canaries in coal mines other than as ongoing chemical monitors, surrogates for chemical analysts. The procedure worked because previous observations had shown that canaries had appropriate sensitivity to particular gases. Another illustration is the recent Ames test for the presence of carcinogens. Chemicals that increase the mutation rate of bacteria are more likely to increase rates of cancer in vertebrates, since both processes are results of damage to DNA. Thus, chemicals that are likely to be carcinogens can quickly and easily be distinguished from others. Many natural compounds are potentially carcinogenic. There is considerable controversy over how to react to the presence of man-made and natural carcinogens at various concentrations, but there is very little controversy about the bacteria's ability to respond in a consistent way to the chemicals.

The significance of particular chemical concentrations for ecosystems is a difficult problem whose solution involves the full panoply of ecological analysis, not merely ecotoxicology. There is no simple way to use toxicological surrogates for important species or for entire communities in a way that will stand up to legal arguments. If it is not convenient to get bass, minnows cannot be used as surrogate bass. However, if organisms have been carefully chosen and pretested for consistency of response, they may well serve as surrogate chemists rather than being chosen as surrogates for the natural ecosystem itself, and debate and litigation may thereby be reduced.

In short, debate and uncertainty still exist about how to assess whether and to what extent damage has been done to an ecological community. In fact, no single species or subset of the total species present is a reliable surrogate. Damage to an organism or microcosm can only be translated into clear assertions about damage to the ecosystem itself if there exists a body of ecological knowledge and theory that cannot be provided by the toxicological data alone. These data and theory do not exist, but there should be sufficient funds for research in this area. At present, university population biologists and ecologists are very grossly underfunded.

CONCLUSIONS

Applied ecology is difficult but not impossible. Action has to be taken, but the problems cannot be solved by off-the-shelf answers. Solutions will require intellectual and empirical depth well beyond what is now available, as well as commitment, money, organization, and work. Most significantly, applied ecology requires rethinking the basis of how ecological problems and their solutions are approached. It is almost too late to start, but tomorrow is even later.

REFERENCES

1. Botkin, D. B., B. Maguire, B. Moore, H. J. Morowitz, III, and L. B. Slobodkin. "A Foundation for Ecological Theory," *Mem. Inst. Ital. Idreobiol. Suppl.* 37:13–31 (1979).
2. Slobodkin, L. B. "Intellectual Problems of Applied Ecology," *BioScience* 38:337–342 (1988).
3. Sloof, W. *Biological Effects of Chemical Pollutants in the Aquatic Environment and Their Indicative Value* (Utrecht: Drukkerij Elinkwijk BV, 1983).

CHAPTER 7

Clean-Up of Contaminated Lands: How Clean is Clean Enough?

Jerry J. Cohen

INTRODUCTION

Recent events related to the Exxon Valdez oil spill in Alaska illustrate the problem of clean-up of contaminated lands. In August 1989, after 5 months of clean-up effort costing over $0.5 billion, the Exxon Corporation announced its plan to complete the effort within another 2 months and before the onset of winter. This announcement was greeted in the media by a flurry of statements from various government agencies and "public interest" groups to the effect that: (1) total clean-up must be assured and (2) the party causing the problem should not be trusted to police itself by determining the sufficiency of clean-up. In the acrimonious discussions that followed, a very important point seemed to be missing from all the media coverage: there are no clear and definitive criteria or guidelines against which the adequacy of the clean-up effort could be evaluated. Under current policies, the question of "how clean is clean enough" is a largely subjective determination. There are no clear targets at which to aim. Although it is commonly understood that clean-up of every last molecule of contamination is not physically possible, characterization of definitive conditions that define clean-up adequacy has yet to be accomplished.

Development of definitive criteria for clean-up of contaminated lands requires, as a first step, the determination of clear objectives. To answer the question "how clean is clean enough?", one must first decide the answer to "clean enough for what?"

POSSIBLE CLEAN-UP OBJECTIVES

Several possible objectives for clean-up operations have been either explicitly stated in justifying clean-up activities or may be implicitly inferred from an assessment of plans and operations. Possible rationale for clean-up are discussed in the following.

Protection of Human Health and Safety

The most frequently expressed reason for clean-up of contaminated lands has been the desire to prevent exposure of present and future populations to the toxic contaminants. The presumption, of course, is that such prevention is a required or, at least, an important factor in protection of health. Certainly, in the public mind, such laws as RCRA and CERCLA are needed to protect its health and safety. Surely this was the prevailing belief in Congress when this legislation was passed. However, from the standpoint of science and logic, there is a serious question as to whether implementation of laws requiring remediation of contaminated lands will, in fact, contribute in any significant way toward protection of public health. That question is a major theme for this discussion.

Compliance with Legal Requirements

Regardless of its implications to health and safety, the law is the law and therefore it must be obeyed. However, since many provisions of recent environmental laws and regulations tend to be rather vague, and therefore subject to interpretation, it is often difficult, if not impossible, to quantitatively assess how and when compliance has been achieved. The USEPA Hazardous Waste Ranking System (HRS), for example, provides a means of ranking, or prioritizing, hazardous waste sites as to their relative hazard, but provides no guidance as to how low on the HRS scale a site must rank to be considered "clean." It should be noted that the HRS is itself highly subjective in nature and therefore difficult to apply in a uniform and consistent manner.

In any case, the concepts of "compliance with legal requirements" and "protection of public health" are not synonymous in nature and, in fact, may bear little relation to each other.

Undetectability of Hazardous Contaminants

Certainly compliance with the law could be considered if the total absence of hazardous materials could be proved. However, this approach involves questions of limits of detectability, adequacy of sampling methods, and the general requirement for "proving a negative," which is a logical impossibility. It should also be noted that for any given contaminant, any correlation between levels of detectability and level of hazard would be purely coincidental, since those material properties that contribute to detectability and those contributing to hazard are largely unrelated to each other.

No Increase in Background Levels

Some regulations call for "no degradation" of the environment. Such requirements may be reasonably interpreted to mean no increase in background levels of hazardous materials. Since background levels are highly variant, it may be difficult to determine which level of background most reasonably provides the basis for comparison. Certain areas of the United States have extremely high levels of naturally occurring toxic minerals (some surface waters in various playa regions in the western states can actually contain lethal levels of contaminants such as arsenic, boron, and selenium due to entirely natural conditions), while in other areas, natural concentrations of toxic minerals may be several orders of magnitude below the level that could constitute any significant risk.

Satisfaction of a Concerned Public

Whatever the nature of the problem related to land contamination, the specter of contaminated land elicits a high degree of public concern. Love Canal and Times Beach are but two examples where a high degree of fear, sometimes bordering on panic, toward toxic contamination has surfaced. Usually, the fear prevails despite the absence of epidemiological evidence that physical harm has resulted or would occur in the future.

Since the public has little technical understanding of the factors affecting environmental movement and toxicology of contaminants, the simple identification of contaminated areas is generally sufficient to stimulate apprehension. If accompanied by anecdotal evidence of harmful effects to residents in the vicinity, this fear is enhanced. Since no community is entirely free from the occurrence of cancer, birth defects, or other harmful effects that could be attributable to toxic waste, such anecdotal evidence is usually not difficult to obtain.

Given that public fear has been established, the question of determining clean-up levels sufficient to assuage concerns becomes intractable. When fear is not based on technological understanding, it is doubtful it could be ameliorated by technical fixes. Large-scale conspicuous clean-up activities might only serve to confirm and aggravate concerns. This is particularly true when trust is lacking for those government and/or industry personnel responsible for the clean-up operation. The actual quantitatively determined level of contamination and its resultant risk is unlikely to affect the level of public concern. Public concern is more likely based upon the effectiveness of public relations efforts and can be highly dependent upon the manner in which the media decide to handle the situation.

Other Possible Clean-Up Objectives

Other possible clean-up objectives could include protection of wildlife habitats, abatement of fire and explosive hazards, and improvement of aesthetic conditions. The latter would include removal of conspicuous drums and miscellaneous junk that constitute visual "pollution," and elimination of offensive odors.

THE NATURE OF THE PROBLEM

An old adage states that "If you don't know where you are going, it's hard to know when you get there." Such is the nature of the problem of meeting current environmental requirements. The spate of environmental legislation produced in recent decades (RCRA, TOSCA, OSHA, CERCLA, SARA, etc.) could be among the most onerous to be foisted upon the American public. The technical objectives of this legislation are often unclear, and the regulations stemming from the laws tend to be inconsistent, and confusing.[1,2] Although the rationale, either explicit or implicit, relates to protection of public health, several assessments indicate the degree of health protection that might be afforded would be marginal at best.[3-10] A literature search by this author for scientific (i.e., nonanecdotal) evidence to support the presumption that contaminated land areas could pose a significant health threat was largely unproductive. Perhaps the most comprehensive study supporting the view that land contamination from hazardous waste is a significant health problem ("our number one environmental crisis") may be found in the work of Epstein and colleagues.[11] Although this work cites no epidemiological evidence to support this view, it does provide a historical summary, along with an abundance of anecdotal citations. The magnitude of the problem is described in terms of quantity, i.e., "80 billion pounds of hazardous waste is produced annually – about 350 pounds for every inhabitant of the United States." It is difficult to understand how the mass (tons, pounds, grams, etc.) per se could translate into level of hazard. It might be noted, for example, that over 1400 billion pounds of coal are produced in the United States annually. This coal contains known carcinogens, as well as significant quantities of radioactivity.

Whatever the nature and degree of risk posed by "improper" disposal of hazardous waste, the cost for clean-up of contaminated lands is likely to be enormous. For example, recent cost estimates for the clean-up of United States Defense facilities alone amount to over $200 billion.[12,13] This figure approximates the estimated cost for the Savings and Loan "bailout," over which Congress is currently agonizing. It also approximates the entire monetary cost for conducting the Vietnam War. Of course, the cost for clean-up of contaminated lands is strongly dependent upon the extent of clean-up required. In the absence of clear and definitive clean-up standards, it should be noted that clean-up cost estimates must be considered highly tenuous.

If health protection were the objective, it might be reasonable to ask whether this expenditure of a significant fraction of our national wealth will be cost effective. If, as previous risk assessments indicated, the threat to human health is minimal to begin with, then the marginal improvement would, at best, be very small, and the marginal return per unit cost would be minuscule.

HISTORICAL ASSESSMENT

The previous discussion suggests that current laws and regulations for clean-up of contaminated lands might constitute an example of gross overkill where the results gained in improved health and safety would not justify the vast expense involved. If this conclusion is accepted, it might be reasonable to inquire as to how this state of affairs came to be.

In ancient times and through the Dark Ages, the understanding of natural phenomena, and particularly the practice of the healing arts, was largely based upon superstition (by today's standards). A prevailing belief during that time was that the stars and planets controlled all parts of the human body (astrological determinism). During the 15th and 16th centuries, such beliefs were slowly replaced by the scientific approach. A major figure in stimulating this change was the German physician Paracelsus (1493 to 1541), who created the beginnings of modern toxicology and pharmacology. From his observations of the effects of various natural and man-made substances, he wrote "What is not a poison? All things are poisons and none without poison. The dose alone makes a poison."[14] This Paracelsian philosophy became the generally accepted paradigm in toxicology and medicine in subsequent centuries and persisted up to the 1960s when it was apparently replaced by a new paradigm.

This new paradigm could be characterized by the assumption that any material shown to be harmful at high doses should also be considered harmful, to some degree, at all dose levels greater than zero and that any substance shown to be harmful to one species of animal could also be considered harmful to all other species, including humans. In other words, the assumption is that there is no threshold of dose below which harmful effects will not occur and that it is valid to infer low-dose effects by extrapolation from observed effects at high-dose levels and/or effects in other species. The new paradigm or "new toxicology" as described by Efron[15] was based on the moral imperative that society must protect the most susceptible of its members from any degree of risk. Its application was primarily intended to provide protection against carcinogens as well as "suspected" carcinogens. As described by Efron, "The 'new toxicology' was, in sum, a demand for a new moral-political approach to toxicology which would, with the aid of sociologists, lawyers, and economists, rationalize the pursuit of suspected industrial mass murder without the need for the empirical evidence of harm to man that is demanded by science."

Acceptance of the new toxicology, plus the pervasive fear of cancer and belief that most cancer and other disease is of environmental origin, provided much of the impetus for developing the profusion of environmental legislation in recent decades.[10]

Regardless of how well intended this legislation may have been, it nonetheless appears to be illogical, unscientific, and in the long run, could be detrimental to man's overall health and safety. It rejects the traditional scientific concepts of proof based upon data, analysis, logic, reproducibility, and peer review. Instead, it insists upon total and demonstrated safety. Although something can be proven dangerous, proving it is safe is impossible.

DISCUSSION

Considering such problems, it is no wonder that recent environmental legislation has resulted in several decisions that can be considered curious, if not totally absurd. A few are

• Many common foods are exempted from regulation despite the fact that their carcinogen content would otherwise require their management as toxic materials.[16] If they were wastes, they would qualify as hazardous wastes. When asked why warnings were not required for natural carcinogens in food, a federal official responded, "Such warnings would be so numerous they would confuse the public . . . and not enhance public health."[17]

• Saccharin must be treated as a hazardous waste, yet it is allowed as a food additive in the United States.

• Although saccharins are legal in the United States, cyclamates are banned. In Canada, saccharins are banned while cyclamates are permitted.[18]

• A study by the Office of Technology Assessment[2] indicates that even if current requirements are followed exactly, many hazardous waste disposal facilities in compliance with RCRA provisions would likely become Superfund sites at some time in the future. In a discussion of inconsistencies between the two laws, Hileman[1] expressed puzzlement over "whether the writers of the RCRA regulations worked for the same government as the authors of the superfund regulations."

The contention that recent environmental laws are likely to prove inimical to public health may seem peculiar. It might be argued that even if they do no good, at least they will do no harm. After all, "you can't be too safe," or can you? To understand how these laws could actually prove detrimental to public health in the long run, it might be best to take an economic viewpoint.

In all societies, including the United States, there are more good ideas than there are resources with which to execute them. Given that societal resources are limited, it is prudent to plan, prioritize, and allocate these limited resources where they will do the most good (i.e., "get the biggest bang for the buck"). This is also true when it comes to health and safety. It is neither immoral nor irrational to spend the limited resources that have been allocated toward the protection of public health in the most effective manner. In the absence of scientific evidence that contaminated lands pose any significant threat to present or future public health, it is hard to conceive that the planned clean-up of these areas would constitute anything other than a needless squandering of limited public resources. The vast clean-up efforts planned will necessarily divert those resources from other areas of health and safety where they might have been far more beneficially applied (e.g., traffic safety, cancer research, narcotics prevention, etc.). As a result, the overall public health must suffer.

SUMMARY

An assessment of recent legislation such as RCRA and CERCLA leads to the impression that political considerations may have provided a greater motivation in legislative development than did science. Scientists may be poorly qualified to advise on standards for clean-up of contaminated lands and may require the counsel of a lawyer or politician to know "how clean is clean enough?"

REFERENCES

1. Hileman, B. "ES&T Outlook: RCRA Groundwater Protection Standards," *Environ. Sci. Tech.* 18(9):282A (1984).
2. *Groundwater Protection Standards for Hazardous Land Disposal Facilities: Will They Prevent More Superfund Sites?* OTA Memorandum (Washington, D.C.: U.S. Office of Technology Assessment, 1984).
3. Bovard, J. "Some Waste Clean-up Rules are a Waste of Resources," *Wall Street Journal,* February 15, 1989.
4. Efron, E. *The Apocalytics* (New York: Simon & Schuster, 1984).
5. Cone, M. "Scoffing Through the Apocalypse," *Calif. Mag.* February:75-77,127 (1986).
6. "Dioxin Report: A C&EN Special Issue," *Chem. Eng. News* June 6 (1983).
7. Smith, R. J. "Love Canal Study Attracts Criticism," *Science* 217(20):714-715 (1982).
8. Kolata, G. B. "Love Canal: False Alarm Caused by Botched Study," *Science* 28:1239-1240 (1980).
9. *Love Canal: The Facts (1892-1980),* Factline No. 11, Informational Pamphlet (Houston: Hooker Chemical Co., 1980).
10. Sagan, L. *The Health of Nations,* (New York: Basic Books, 1987).
11. Epstein, S. S., L. O. Brown, and C. Pope. *Hazardous Wastes in America* (San Francisco: Sierra Club Books, 1982).
12. Beam, P. M., M. J. Carricato, W. H. Parker, and A. Talts. *Department of Defense Restoration Program, Waste Management, '89* (Tucson, AZ: University of Arizona, 1989).
13. Lehr, J. C., L. D. Eyman, and W. W. Thompson. *Department of Energy Defense Programs Environmental Restoration Program Update, Waste Management, '89* (Tucson, AZ: University of Arizona, 1989).
14. Clayton, G. D., and F. E. Clayton. *Patty's Industrial Hygiene and Toxicology* (New York: John Wiley & Sons, 1977).
15. Efron, E. "Behind the Cancer Terror," *Reason* May:22-30 (1984).
16. Ames, B. N. "Water Pollution, Pesticide Residues, and Cancer," *Water* 27:23-24 (1986).
17. Lutzker, L. G. "Making the World Safe for Chicken Little, or the Risks of Risk Aversion," in *Low Level Radioactive Waste: Science, Politics, and Fear* (Chelsea, MI: Lewis Publishers, 1988), pp. 176-191.
18. Whipple, C. "Redistributing Risk," *Regulation* May-June, 1985.

CHAPTER 8

The Savannah River Site as a National Environmental Park

Eugene P. Odum

INTRODUCTION

Since the beginning in 1951, the Savannah River Site (SRS) has had three missions: (1) production of materials for nuclear weapons, (2) production of forest products, and (3) environmental research. The latter had a modest beginning, with small grants to the Universities of Georgia and South Carolina to begin inventories of flora and fauna, and to the Philadelphia Academy to start species diversity inventories of the Savannah River itself directed by Ruth Patrick. The United States Public Health Service also funded a 3-year study of the Savannah River with emphasis on water chemistry and fish for establishing descriptive baselines for assessing future change.

It soon became apparent that unusual opportunities were available for research on thermal change and flooding resulting from the one-way flow of water through the reactors, the fate and effect of radioactive materials in natural systems, and the use of radionuclide tracers for elucidating ecological processes. The large area of natural environment not directly affected by plant operations also provided excellent opportunities for more functional and long-term studies of basic ecological processes, such as ecological succession, nutrient cycling, and energetics, not to mention the study of wildlife species and special habitats such as wetlands. The Atomic Energy Commission (as the Department of Energy [DOE] was called in those days) established a permanent laboratory with a resident staff on the site in 1960 in order to take advantage of these opportunities.

Support for environmental research has increased steadily. Direct support from DOE (and its predecessors) for the Savannah River Ecology Laboratory (SREL) passed $50 million in 1988, and this does not include numerous grants to individual researchers from other federal agencies and universities, support for work of the Savannah River Laboratory (SRL) operated by the prime contractor, and the United States Forest Service research. Publications from SREL alone now number more than 1400.

The importance of environmental research was officially recognized in 1972 when the SRS was designated the first "National Environmental Research Park" (NERP). At that time, there was some pressure within Congress and the federal government to return some of the SRS to private ownership on the assumption that peripheral areas were no longer needed for security. Since long-term study plots had already been established in these peripheral areas, the concept of an "outdoor laboratory" or "research park" designation for land associated with United States national laboratories was developed within the Atomic Energy Commission. A memo by John Trotter, then Director of the Division of Biology and Medicine, recommending such a status for the SRS was approved on April 5, 1972. Thus, the value of the land for research was the basis of the designation rather than for security that the boundaries of SRS remained essentially intact. In subsequent years, site reviews have been favorable. For example, a review in 1987 by Liz Goodson and William Osborn concluded that "Since its designation in 1972 as the first NERP, SRP has demonstrated excellent progress towards realization of the long-term potential of the NERP concept." The reviewers also commented on the practical problems of conflicts in land/resource management.

SOME NEGATIVES

In a commentary[1] prepared for a University of Georgia newsletter, I wrote:

> "To achieve the lofty purpose for which NERP was founded there must be (1) a strong effort to establish communication between academia, industry, government and the public, all of which have different priorities and "vested" interests in the facility; and (2) a unifying concept, goal or focus towards which different lines of research can be directed. We think we have such an overall guiding principle in our concept of *Ecosystem Management,* by which we mean the application of ecological principles to the developmental planning and management of man and environment as a whole."

Unfortunately, communication between the SRS, Washington, and the public has been far from desirable, but with new management and a renewed national and global concern for the environment, perhaps communication will improve.

I once had high hopes that the SRS could become a model for an industrial park in which potentially hazardous operations would be buffered by large areas of uninhabited natural environment. A colleague and I even wrote on this idea:[2]

"The concept of the power park is not really new. Man has always depended on natural areas adjacent to his cities and factories to assimilate the end products of civilization's metabolism. What is new perhaps is the idea of setting aside natural areas free from human habitation for this purpose and establishing a reciprocal design coupling the waste maker and the waste assimilator into an integrated or symbiotic system. Such a strategy creates an economically attractive way to use the solar-powered free work of nature to ameliorate the disorder inherent in man's energy-conversion processes. Although the Savannah River Plant reservation was originally established solely to provide security for the production of weapon material, it turns out that the large size and high natural diversity of the area make it an ideal location for basic research and experimentation to determine how the coupling can be accomplished without undue stress to the biotic components of the living filter."

Unfortunately, I would not at this time recommend the SRS as a model for a buffered industrial park. In retrospect, the problem was that the large area of buffer made it tempting for plant operations (the contractor) and the production foresters to keep nibbling away at it. In the beginning, the plan was to leave the entire Upper Three Runs watershed and the mile-wide Savannah River floodplain (totaling about one quarter of the area) completely undisturbed as a buffer and "control" for the rest of the area that would be modified and impacted in various ways. In the absence of an agreed upon and written down land-use plan, these natural areas have been gradually encroached upon for waste or thermal discharge as the pressure for ever more weapons material mounted (especially during the Reagan administration). Likewise, foresters were tempted to undertake some kind of "stand improvement" management in natural old growth. In the case of Upper Three Runs, clearcutting and bulldozing endangered a national treasury of unusual and rare aquatic species even before they could be inventoried!

The trouble with "management" is that it costs money, and once "managing" begins to create an unnatural forest, "managing" (i.e., spending money) must keep it in an unnatural condition (like maintaining pines on the floodplain). The ideal natural area buffer in one that is self-maintaining and does not require any money to maintain its assimilative capacity. Finally, the concept of "reciprocal design" (i.e., where waste is reduced to the level and type that can be assimilated by the ecosystems present) has never entered the picture.

At least twice in the past 15 years, a meeting has been convened with the prime contractor, the United States Forest Service, and SREL to discuss the need for a land-use plan. Participants present at these meetings agreed that a written down and agreed upon land-use plan was necessary, but after the meeting, the recommendation was not followed up. It got lost in the bureaucracy, or maybe it was because the Department of Defense (DOD) did not want any limitations (or at least it wanted the option of a "negotiated" limitation) on what might be done in the future to maintain the "defense effort." Setting up a buffered industrial park requires making a land-use plan at the beginning and sticking to it!

SOME POSITIVES

The SRS has had bad publicity lately with the aging reactors and the accumulation of waste products. However, accomplishments have been achieved with environment research on the SRS.

Thermal Effects

When temperature increases but remains within vital limits (as in most of Par Pond), cold-blooded animals benefit first. Alligators like it, and fish grow faster and larger; however, the warm water also favors fish diseases, so mortality is high. With so much "dead" meat, turtles switch from plant-eaters to meat-eaters and grow to great size. Thus, thermal pertubation can affect the food web in unexpected ways. If global warming does indeed occur, then warm-blooded humans are going to be faced with more competition from cold-blooded types!

Much has been learned about "pulsed systems" where ponds and streams have been subjected to alternate heating and cooling as reactors are turned on and off. Heat-tolerant species of phytoplankton and zooplankton that are very rare at ambient temperatures quickly become dominant when the water is warmed, a dramatic illustration of the importance of "hidden" biodiversity when it comes to adapting to changed conditions.

Ecological Succession

With agriculture suddenly abandoned on about one third of the land in 1951, an unusual opportunity was available to study and experiment with ecological succession. From a functional standpoint, succession on the old fields involved a series of steady states associated with a change in major life forms, rather than a continuous change associated with species change as is usually postulated. On the sandy soils of the area, nutrients are leached rapidly (as in the tropics) beyond the reach of pioneer forbs and grasses. Productivity then declines until the next life form, pines, come in and recycle nutrients with deeper roots and mycorrhizal symbionts. Some very important research with the latter, directed by Don Marx of the Athens Forest Service Laboratory, was carried out at the SRS. Heavy inoculation of pine seedlings with an early successional species of mycorrhizae enables pines to grow in "borrow pits" where all soil has been removed for road building.

The Delta

When heated flood waters were released in Steel Creek, subsequent erosion created a large delta at the point where the creek enters the Savannah River Floodplain. The delta became a temporary feeding ground for the endangered wood stork and, when the flooding stopped, an ideal place for the study of the regrowth of cypress-gum and other swamp forests. Some sophisticated remote sensing made it possible to follow the succession in great detail and to determine the conditions favorable for regeneration of cypress and other valuable hardwoods. More than 50 papers have been published as results of work on the Delta. Wetlands are now one of the three "departments" at SREL (the others are "Biogeochemistry" and "Wildlife and Stress Ecology").

The 3-412 Carolina Bay

For many years now a continuously monitored "drift fence" has been maintained around this wet depression through wet years and dry years. The "traffic" of amphibians and reptiles in and out of this area is unbelievable! Again, this is an example of a project that can have meaning only if conducted over a long period of time in a large reservation. Incidentally, turtles are not usually considered particularly mobile, yet some that got into waste ponds and picked up some radiocesium and radiostrontium were found to move some distance outside the SRS, which is why as big a buffer as possible is needed.

Radionuclides Tracers

In the early days, quite a bit of research was done on the use of short halflife radionuclides, such as ^{32}P, ^{65}Zn, and ^{45}Ca, to chart movement of both nutrients and species in intact ecosystems. Tagging plants allowed investigators to "isolate food chains." When two co-dominants in old-field succession, *Leptilon* (horseweed) and *Heterotheca* (camporweed), were tagged in separate quadrants, the former supported a large number of ants that feed on nector glands but not many other insects, while the other equally successful species supported a much larger variety of insects. These are contrasting methods of survival: provide food for predators that keep the herbivores off or produce "anti-herbivore" chemicals that reduce herbivory. In recent years, stable isotopes used as tracers have largely replaced radioactive ones, thus eliminating the possibility of radiation effects.

Radiation Effects

By using both fixed and portable gamma sources to radiate both organisms and communities, it was discovered that wild species of both plants and animals differed from domestic or laboratory species in their response or tolerance to ionizing radiation. For example, the LD_{50} for native rats and mice was higher than that for laboratory white rats. On the other hand, an irradiated whole pine forest was slow to recover.

Plutonium Studies

If a choice could be made, plutonium would probably not be utilized or studied. However, if a plutonium containing satellite at Cape Canaveral blew up and contaminated nearby orange groves, then something about the fate and cycling of this hazardous element should be known. Perhaps fortunately, once trans-uranics are deposited in sediments and soils, they become relatively immobile.

Migratory Birds

The largest possible research park still could not keep migratory birds from taking materials out. Since birds are known to concentrate certain nuclides such as ^{90}Sr and ^{137}Cs, it is important to monitor body burdens, particularly in waterfowl. A good deal of attention has been given to waterfowl study at the SRS. So far, removal of fission products from the area is minimal, but it should be watched. As a result, much is learned about the movement and behavior of the birds themselves.

An interesting sidelight was discovered about hummingbirds. One year, a premigratory gathering in a patch of Crotalaria in an old field enabled investigators to measure fat deposition just before migration. This research proved for the first time that these birds accumulate enough fat to fly nonstop across the Gulf of Mexico.

Mesocosms

Surrounding the new, modern laboratory, dedicated in 1977, are acres of "mesocosms," replicated outdoor enclosures where hypotheses based on field observations are tested under replicated semi-controlled conditions. These mesocosms range from "rhizatrons" for study of root development in wetland species, to cattle tanks for aquatic studies, to fenced enclosures for terrestrial populations, and to aviaries for bird behavior studies.

Incidentally, the first enclosure was built by Robert Norris in 1958 as an aid in his studies of Savannah sparrows. This research won him the Ecological Society of America's coveted Mercer Award for the best paper published in that year. Thus, the mesocosm has been a tradition at SREL.

Taxonomic Studies

In recent years, DOE has provided grants to bring leading taxonomists to the site to prepare annotated lists and keys of all species that have been identified for important but not well-known groups of organisms such as ants, aquatic insects, grasses, etc. These studies have been published in a series of NERP bulletins.

These are just a very few of the highlights of nearly 40 years of research at the SRS. Recent emphasis at SREL has been on the genetics of natural populations that are under different kinds of stress (as, for example, hunting pressure in deer) and on biodiversity in microbial populations.

REFERENCES

1. Odum, E. P. "Commentary," *Res. Rep.* 7(1):8–9 (1972).

2. Odum, E. P., and R. Kroodsma. "The Power Park Concept: Ameliorating Man's Disorder with Nature's Order," in *Thermal Ecology II,* G. W. Esch and R. W. McFarlane, Eds. (Springfield, VA: National Technical Information Service, 1977), pp. 1-9.

CHAPTER 9

A Strategy for the Long-Term Management of the Savannah River Site Lands

F. Ward Whicker

INTRODUCTION

When the Savannah River Site (SRS) was created nearly 40 years ago, its mission was clearly defined. The mission was considered of such national importance and urgency that the pragmatic matters of having a sufficiently large land area and adequate reactor cooling water probably took precedence over all others. The question of the land management strategy in 30 to 40 years was probably given little thought. The principal function of the site, namely production of nuclear materials for military purposes, was the overriding goal, especially during the first 2 or 3 decades of operation. Exempt from today's stringent environmental regulations, the Savannah River Plant, like other facilities in the country's weapons production complex, used relatively expedient operating and waste disposal practices. Urgency of production, coupled with a general lack of waste management research and experience, no doubt led to some "seat of the pants" judgments in the early years of operations.

Given this history, it seems remarkable that the safety record and ecological condition of the SRS is as good as it is. Nevertheless, many individual sites within the SRS have been measurably impacted with physical insults, hot water, and radioactive and chemical contaminants. These impacts are largely a legacy. Today's technology, regulatory constraints, and managerial attitudes have created dramatic changes in operations of the SRS and other weapons complex facilities.

A major challenge for the Department of Energy (DOE) at present is how to best manage the lands and facilities under its care. The challenge is particularly complex now because of the deteriorated condition of many of the reactors and associated facilities, intense public concern, and uncertainty about the quantities of nuclear material that may be needed in the future.[1] Regulatory requirements, often driven more by public opinion and politics than by science and fact, are confusing, sometimes conflicting, and still evolving. Public support and DOEs general credibility have probably never been lower. Yet, in the face of this gloom, prospects for nuclear disarmament have never seemed brighter.

In response to this situation, Admiral James Watkins, the new Secretary of Energy, has officially pronounced that matters of health, safety, and environment will now take precedence over nuclear materials production.[2] A high priority will be cleanup and restoration of contaminated sites, as well as modernization of the facilities used to produce nuclear materials. Much more careful and technologically based waste management strategies will be sought for both present wastes and those to be generated in the future. To accomplish these objectives, changes in funding priorities and in the weapons complex "culture" are clearly necessary. The bill for cleanup alone has been estimated to exceed $100 billion,[3,4] a figure nearly tenfold DOEs current total annual budget. The time required for cleanup is difficult to estimate because cleanup criteria, standards, and technologies need more development, and the process could easily require 1 or 2 decades or longer.

The characterization and evaluation of contaminated sites at SRS and other DOE sites will likely be pursued actively in the near future. Cleanup standards should be based on actual risk to human health and environmental quality. Spending large sums to clean up contamination at levels that pose no significant risk by remaining in place does not make sense. However, if credible, scientifically sound risk assessments are not conducted and sold to the public and the regulatory community, cleanup standards are likely to be driven by public fear and ignorance to zero or near zero levels of residual contamination. This will not only escalate cleanup costs but also the ecological impacts resulting from application of the cleanup technology.

CURRENT STATE OF THE ENVIRONMENT AT THE SRS

The SRS encompasses a roughly circular area of about 300 mi^2 (>800 km^2). It is the largest contiguous federal land holding of its type east of the Mississippi River. Over 20 km of the SRS is bordered by the Savannah River; the remainder is bordered primarily by rural farmland and pine plantations. The site has many natural biotic communities characteristic of the region, including lowland hardwood forests, turkey oak-long leaf pine associations, cypress-tupelo swamps, blackwater streams, and Carolina bays.[5] Numerous habitats influenced by human activities also exist on the site, including old fields, large pine plantations, abandoned farm ponds, several large reservoirs, canals, and artificial seepage basins. Roughly ten large industrial sites (reactors, separations facilities, waste treatment and storage areas, and laboratory areas) and numerous smaller, developed sites are scattered across the SRS. These sites are connected by a network of roads, rail lines, power lines, and other types of corridors.

The placement of man-made structures, combined with the natural topography, has created a mosaic of numerous community types, giving the site a high degree of pattern diversity.[5] Most of the community types are fragmented into numerous individual stands, each similar to one another but separated by other community types or man-made corridors. Several streams and their associated riparian habitats cut across large sections of the SRS and then enter the Savannah River. The cypress-tupelo swamp is largely contiguous along the Savannah River and lower reaches of tributary streams.

Protected for approximately 4 decades from public use, the majority of the SRS (I subjectively estimate >80% of the land area) has become a major sanctuary for many species of plants, fish, and wildlife. Much of the area has been relatively unexploited, although forestry and big game hunting operations exist. In 1987, for example, some 27 million board feet of timber were produced, and nearly 900 ha were planted with pine seedlings.[6] Controlled public hunts produced >600 deer and >100 wild hogs in 1987.[6] The SRS may also be of importance for the protection of rare or endangered species, such as wood storks, red cockaded woodpeckers, bald eagles, and American alligators.

From an ecological viewpoint, the SRS is undoubtedly one of the best characterized, most thoroughly studied areas of its size in the world. The University of Georgia's Savannah River Ecology Laboratory has had an active and growing research program at the SRS for over 30 years. This program involves both basic and applied research on the aquatic and terrestrial ecosystems on and near the site. In 1972, the site was officially designated a National Environmental Research Park. With this status, outside researchers from academia and other endeavors were encouraged to conduct studies on the site. The prime contractor, Westinghouse (formerly E. I. duPont de Nemours and Company), also has a sizable program on the SRS that is concerned with applied environmental research, most of which is directly relevant to plant operations and regulations.

Although a large fraction of the SRS has been only modestly impacted by human activities, many sites have been altered physically or contaminated by chemical or radioactive materials. Probably the most obvious and significant environmental impacts have been those that physically result from the building of facilities, roads, utility corridors, dams, canals, etc. In addition, hot water discharges from reactor operations have significantly altered sections of some streams and associated riparian vegetation, portions of the cypress-tupelo swamp, and biotic communities of reservoirs used for dispersing heat. Over 200 individual locations within the SRS have been identified that contain possibly significant levels of radioactive and/or chemical contamination. In addition, certain streams, reservoirs, wetlands, terrestrial areas, fish, and wildlife contain detectable (but generally insignificant from a health or ecological risk standpoint) levels of radioactivity or chemical contamination. Also, extensive areas of measurable groundwater contamination exist at the

SRS. Chemical and radioactive contamination on and around the SRS is monitored frequently and extensively by Westinghouse.[6]

ON THE QUESTION OF CLEAN-UP AND MANAGEMENT OF CONTAMINATED SITES

Life has evolved, persisted, and flourished in an environment of natural cosmic radiation, primordial radioactivity, and tens of thousands of natural chemical substances, many of which are poisonous or carcinogenic. In sufficiently elevated concentrations, many of these natural substances would be harmful to life. Radionuclides and chemicals created by man are similar to their naturally occurring counterparts in that they may or may not be harmful, depending on their concentrations in the environment. Modern analytical systems can easily measure radionuclides and chemicals in the environment at levels that may be biologically innocuous. Despite these truths, many people share the misconception that the mere presence of a man-made contaminant in the environment will necessarily cause problems with human health or with plant and animal life.

Public ignorance of the general principles of toxicology, combined with general distrust of DOE and its operations, has created an atmosphere of fear. Emotional feelings from the public have led to statements and decisions in the political and regulatory arenas that have perception and fear as their primary basis. Often lacking the necessary scientific training and research data, and faced with enormous pressure from the public, it is not surprising that decision makers and regulators sometimes act on perception rather than fact.

Given the enormous cost estimates that have been published for cleanup of DOEs contaminated lands and waters, it seems imperative that site-specific standards be developed to determine first whether cleanup is warranted at all. Except for USEPAs drinking water standards, there currently are no standards for cleanup of environmental media. For example, no formal cleanup standards exist for soils or sediments, which generally hold >95% of the total environmental inventory of most contaminants. The only scientifically objective method of establishing such standards is to conduct a rigorous, site-specific risk assessment that is based on the best-available models and data. Such a risk assessment can establish the relationship between levels of environmental contamination and health risk to people or impacts on plants and animals living on the contaminated site. The risk assessment models, of course, must be based on good science and thoroughly tested if they are to achieve any measure of credibility. If this approach is eventually used to determine cleanup criteria, it will require an investment of time and resources to develop and test credible models. The investment will pay for itself many times over compared to that of unnecessary cleanup and restoration, but the time commitment will necessarily delay many cleanup decisions for perhaps a decade or more. This general approach does not preclude more immediate cleanup of highly contaminated sites that do not require a rigorous risk assessment to determine the necessity of remedial action.

Another benefit of a deliberate and scientifically based approach to the question of cleanup criteria is that it buys time to develop more efficient and perhaps less ecologically damaging methods of cleanup and restoration. While promising technological methods of soil and water decontamination are under development currently, the ecological impacts of full deployment of such methods are not well established. For example, water and vapor extraction from the ground, introduction of microorganisms that degrade organic compounds, in situ leaching or vitrification, etc. all carry the potential for unanticipated ecological impacts. The question "Is the cure worse than the disease?" must be asked. Finding the answer to this question will undoubtedly require time for appropriate research.

Many sites within the SRS boundaries contain measurable levels of long-lived radionuclides in soil or sediment. ^{137}Cs and ^{239}Pu are common examples. These radionuclides tend to absorb very strongly to soils and sediments; thus, it is doubtful that anything short of physical removal of the material can be effective in removing the contamination. Physical removal of soil or sediment, especially when large areas are being reclaimed, is not only expensive but ecologically disruptive. Furthermore, there remains the question of what to do with the contaminated soil once it is removed.

One example of this type of situation is the sediment in Pond B, which was contaminated with substantial quantities of radioactivity from R-reactor in the early 1960s.[7] After >20 years of biological succession and nonexploitation, Pond B is biologically diverse and its use as a public fishery might be considered at some point. Over 99% of the total radioactivity is contained in the upper 15 cm of the sediments, and this provides a reservoir of ^{137}Cs that will maintain measurable levels of the radionuclide in fish and other wildlife for at least several decades. For this reason, the public and regulatory agencies might push for contaminated sediment removal from the lake prior to its use as a fishery. Calculations indicate, however, that even though the ^{137}Cs levels are measurable, quite high consumption rates of fish (>10^4 g/year) from Pond B could take place without exceeding continuous radiation exposure standards for the public (1 mSv/year).[7] As long as the reservoir is maintained at or near full capacity, external radiation from ^{137}Cs in sediments would be insignificant as a potential source of exposure. The monitoring and ecological cost for allowing controlled, public use of this fishery would be very small compared to that of aggressive cleanup and restoration. This is probably also true for a large fraction of the other contaminated sites at the SRS.

Smaller areas that are contaminated with levels of radioactivity or chemicals in concentrations sufficient to constitute a measurable potential hazard to people, plants, or animals may definitely warrant aggressive remediation. In cases where such areas may be acting as a significant source for further dispersal of the contamination, immediate cleanup using current technology may be justified. On the other hand, highly contaminated sites where the contaminants are immobile probably need immediate protection from intrusion of people, biota, water, etc., but their ultimate cleanup might be postponed until more effective and efficient technologies can be brought to bear on the problem.

It can be argued that some sites, such as those containing underground tanks with large volumes of wastes[8] should remain indefinitely without cleanup. Removal and reprocessing of such wastes would not only be extremely expensive, but possibly risky for workers and the environment as well. One approach is to immobilize such wastes in place and make engineering improvements to provide physical protection and to divert water that could otherwise infiltrate into the waste. Such areas will probably require surveillance and monitoring for indefinite periods of time. This approach carries the risk of loss of institutional control and inadvertent intrusion by innocent parties at some future time, but such a risk may be no greater than the risk of other catastrophes such as war, floods, earthquakes, etc. This risk can also be minimized with adequate markers, warnings, documentation, etc. If this line of reasoning ultimately prevails, it is clear that at least some portions of the SRS will remain dedicated for the purpose of hazardous waste internment for indefinite time periods.

Assuming the scenario of indefinite hazardous waste internment at the SRS, the question of the appropriate and necessary size of the area dedicated for this purpose will be worthy of serious consideration. Some buffer zone seems prudent and necessary to reduce the risk of migration of wastes from the site or that of inadvertent intrusion from the outside. Groundwater flow patterns, site geology, contaminant adsorption isotherms, etc. are a few of the technical questions that must be answered. Economic and political factors can also be expected to enter into the equation for determining the size of the area to be dedicated for the purpose of indefinite waste internment.

The broad question of clean-up and management of contaminated sites and waste disposal areas is so complex that near-term decisions should be limited to a few cases that are relatively obvious and urgent. A "road map" needs to be developed soon, however, to determine which kinds of research are needed to support decisions that may have major and long-term economic, human health, and ecological implications. This "road map" could take several years to develop. Site characterization, risk assessment model development for cleanup standard setting, and optimal site remediation technology research might require an additional 10 to 20 years before enough knowledge is available to develop and consummate a comprehensive environmental plan for the SRS. Even if the current military support mission of the SRS were to evaporate within the next few years, the site could not be reasonably converted to totally new uses within the next 2 or 3 decades. If the current mission of the SRS continues, the opportunity exists to prevent future problems of environmental contamination and to develop plans and the essential knowledge base for the wise stewardship of the site, irrespective of future changes in its primary mission.

OPTIONS FOR FUTURE MANAGEMENT OF SRS LANDS

Making the assumption that the military purpose of the SRS dissolves at some point and that the site (or most of it) is eventually declared suitable for other uses, several options for its future use are apparent. Such options might include return to the private or public sectors, use as an industrial park, use as a park for environmental or technological research, use as an ecological preserve, or some combination of the above. If some specific mission is identified early enough as being most serving of future local and national needs, planning might significantly facilitate and enhance that mission.

The natural resources of the SRS would clearly be attractive to public investors if the site were to become available. Some could visualize, for example, housing developments, business properties, golf courses, fishing lakes and streams, agriculture, and forestry. Indeed, strong political pressure for such private developments can be expected. An argument against this scenario is the legal and regulatory nightmare that it would likely create. Knowing the past history and contamination legacy of the SRS, future residents or users of the land will find it convenient to blame all sorts of health problems on the government for creating the contamination in the first place and selling the land in the second. Publicity for ensuing legal contests could devalue the land and adversely affect many additional investors, much as in DOEs recent $73 million settlement to residents around the Fernald plant in Ohio.[9]

The very qualities of the SRS that led to its choice as a nuclear materials production site will perhaps qualify it for certain industrial endeavors that require large quantities of water and land. Examples of such industrial uses might include energy production or conversion, large-scale manufacturing, etc. Such industrial uses might be on a scale comparable to or even larger than the historical scale of nuclear materials production at the SRS. For example, global warming, acid rain, and overdependence on foreign oil all argue for a reassessment of civilian nuclear electrical power. The large land area could provide adequate protection of surrounding areas in the event of accidents or pollutant releases. Perhaps the main case with which to argue against some industrial missions for the SRS is that other possible uses for the site might better serve the needs of society at large. The SRS is probably not the only place where large industrial plants can be located, but the site may have other unique attributes that cannot be duplicated elsewhere. However, the site has already proved suitable for nuclear power.

The SRS is no doubt a unique place for the purpose of conducting environmental research. Such research could range from very basic studies in diverse fields such as hydrology, geology, atmospheric sciences, biology, and ecology to applied research dealing with the general questions of evaluating, mitigating, and preventing environmental degradation from resource exploitation, technology applications, and industry. The size and natural diversity of the SRS, the existing base of research, and its proximity to several major research universities make it extremely attractive as a park dedicated to research activities. A precedent has already been established along these lines when the site was designated a National Environmental Research Park (NERP). This current designation has more form than function, since the current NERP budget is <$100,000/year. However, with adequate support, the NERP concept could flourish on the SRS.

Not inconsistent with its use as an environmental research park is the current and potential future function of SRS as a major ecological preserve of the southeastern United States. The large size and protection from public exploitation over the past several decades has created areas of immense ecological value. These areas function as breeding grounds and nurseries that help to replenish plant and animal populations, not only for the SRS, but surrounding areas as well. The SRS has several natural corridors and sanctuaries for fish and wildlife migrations. The natural and cultivated vegetation of the site acts to prevent erosion and to improve water quality of the streams and aquifers of the area. The argument can be made that even the historical nuclear production mission of the SRS has been kinder to the local and adjacent environment than other uses, such as intensive agriculture or urban development might have been. With enhanced protection, the SRS can only increase in ecological value in the future. Continued human population and industrial growth in the region is likely and natural areas, particularly those of significant size, will become even more scarce.

The SRS, because of its history, will likely never be a candidate for a "national park" in the same context as existing national parks. Certain areas will probably always be off-limits to human enjoyment, and other areas will likely require pollutant monitoring for perhaps decades. Nevertheless, substantial portions of the site are completely safe for human visitation and have substantial aesthetic appeal. This is particularly true for the reservoirs, streams, and wetlands. Public support for the ultimate mission of the SRS is clearly desirable and, perhaps, essential. Some form of public use and appreciation of the natural resources of the site may provide one of the best mechanisms for achieving public and political support. Hunting and fishing activities may be compatible with the concepts of using the site as a research park and ecological preserve, as long as such activities are carefully supervised and limited in intensity. Less exploitive activities, such as carefully placed nature trails, boardwalks over wetland areas, bicycle trails, and picnic areas, might also be compatible with the site mission, while generating substantial public appreciation.

A PROPOSED LONG-TERM MANAGEMENT STRATEGY

After considering the more obvious options for the future of the SRS lands, it is clear that legitimate arguments for and against each option can be made. Many individual personal value judgments come into play in the various possible land use scenarios. I freely admit to my own personal biases as an environmental scientist and outdoorsman. My hope, however, is that the ultimate strategy for management of the SRS will evolve from a careful analysis of factual and sound information, as well as an objective appraisal of costs and benefits for society as a whole. This will clearly take some time. In the meantime, however, I will be so bold as to suggest a general plan, fully realizing that it is based on a gut feeling and subject to some personal prejudices. I expect no one to buy these ideas outright, but hope that they may simply help to stimulate some additional thinking and dialogue.

The general plan that I propose points toward the ultimate use of the SRS as a multipurpose park (MPP). This plan assumes *a priori* that the nuclear materials production mission has ceased and that, where necessary, contaminated areas have been cleaned up and restored. The three primary purposes of the MPP would be

- An indefinite storage site for nuclear and other hazardous wastes
- A research park for basic and applied environmental studies
- An ecological preserve

These three areas are mutually compatible on the SRS because of its size and diversity. The use as a waste storage site is not meant to imply that the site should continue to receive wastes generated elsewhere. Once the nuclear materials production mission were completed at the SRS, wastes generated there, as well as contaminated structures and equipment, could be emplaced, entombed, immobilized, and protected with the best-available technology. Thereafter, the waste storage sites would be monitored and protected, but no new wastes would be accepted. The presence of near-surface groundwater at the SRS should be reason enough to discourage its continuing use as a receptacle for nuclear or other hazardous waste.[8]

It is not difficult to visualize the compatibility of environmental research and the ecological preserve concepts on the SRS. Most environmental research activities, if carefully designed, can be accomplished with little or no measurable impact on the environment. In cases where destructive sampling or physical disruption of the ecosystem is necessary, the scale of the impact can be controlled and confined so as to preclude measurable damage to community stands or larger landscape units. Very critical or sensitive ecological sites might warrant complete protection and isolation, even from research activities.

It is more difficult perhaps, to visualize the compatibility of permanent hazardous waste internment with environmental research and preservation. This concept would clearly require that the wastes be adequately confined, immobilized, and protected so as to prevent significant migration in the future. This approach, if done to the specifications of best-available technology, would be expensive, but far less so than retrieving, reprocessing, moving, and storing the waste elsewhere. Current waste storage facilities occupy a relatively small land area at the SRS. This waste storage area is surrounded by a much larger zone that can serve as an effective buffer between the wastes and its surroundings. Small-scale contaminant releases from the waste areas might be expected, but the surrounding natural systems can serve, as eloquently stated by Odum and Kroodsma,[5] as "living filters" that can assimilate and, if necessary, adjust to the contaminants.

The sheer size of the SRS offers large tracts that are unimpacted by wastes currently and into the foreseeable future. These areas are immediately suitable for basic environmental research and ecological preservation. Smaller tracts adjacent to waste storage areas, or those that have received substantial quantities of contaminants in the past, should be monitored and studied. These areas are ideal for research in fields such as ecotoxicology,[10] radioecology,[11] and environmental restoration technology. Such areas have great potential value for the knowledge that they could provide. As discussed by Abelson,[12] scientific research lies at the heart of the problem of learning how to solve many environmental problems.

The MPP would clearly need to be divided into several zones, with each zone serving stated purposes. The zones could include waste storage sites, applied research sites, basic ecological study areas, and areas of total environmental protection. Some areas might also be used for some form of controlled public use for recreation and aesthetic enjoyment. This would require a management structure to establish, monitor, and maintain the integrity and purpose of the various zones.

First and foremost, the MPP concept would need to be sold to the public. The intentions of isolating existing hazardous wastes from the public domain, learning how to better protect and manage dwindling natural resources, and saving those few remaining areas of special ecological value seem noble and straightforward. The public should understand and support such intentions. It is hoped that with adequate public backing, the necessary political support would follow.

REFERENCES

1. Subcommittee on Oversight and Investigations. "Health and Safety at the Department of Energy's Nuclear Weapons Facilities," Proceedings of the 101st Congress, 1st Session, Committee Print 101-H. (Washington, D.C.: U.S. Government Printing Office, 1989).
2. Reid, T. R. "Health, Safety Given Priority at Arms Plants," *The Washington Post* June 17, 1989.
3. Marshall, E. "Savannah River Blues," *Science* 242:363–365 (1988).
4. McGuire, S. A. "Cleanup or Relocation: $128 Billion to Clean Up DOE's Wastes? And They Want to Use My Money?" *Health Phy. Soc. Newsl.* XVII (4):9–10 (1989).
5. Odum, E. P., and R. L. Kroodsma. "The Power Park Concept: Ameliorating Man's Disorder with Nature's Order," in *Thermal Ecology II*, G. W. Esch and R. W. McFarlane, Eds. (Springfield, VA: National Technical Information Service, 1977), CONF-750425.
6. Mikol, S. C., L. T. Burckhalter, J. L. Todd, and D. K. Martin. "U.S. Department of Energy Savannah River Plant Environmental Report for 1987, Vol. 1" (Aiken, SC: Westinghouse Savannah River Company, 1988), DPSPU-88-30-1.
7. Whicker, F. W., J. E. Pinder, III, J. W. Bowling, J. J. Alberts, and I. L. Brisbin., Jr. "Distribution of ^{137}Cs, ^{90}Sr, ^{238}Pu, ^{239}Pu, ^{241}Am, and ^{244}Cm in Pond B, Savannah River Site," (Springfield, VA: National Technical Information Service, 1989), SREL-35/UC-66e.
8. Marshall, E. "The Buried Cost of the Savannah River Plant," *Science*, 233:613–615 (1986).
9. Smith, J., and T. W. Lippman. "U.S. Agrees to Pay $73 Million to Settle Suit at Arms Plant," *The Washington Post* July 1, 1989.
10. Moriarty, F. *Ecotoxicology: The Study of Pollutants in Ecosystems* (London: Academic Press, 1988).

11. Whicker, F. W., and Schultz, V. *Radioecology: Nuclear Energy and the Environment* (Boca Raton, FL: CRC Press, 1982).
12. Abelson, P. H. "Editorial," *Science* 245(4917):449 (1989).

CHAPTER 10

The Role of the Endangered Species Act in the Conservation of Biological Diversity: An Assessment

Larry D. Harris and Peter C. Frederick

INTRODUCTION

The Endangered Species Act (ESA) remains a vital element of conservation programming, but is highly inadequate, in and of itself, to forestall the rapid erosion of biological diversity from the many levels of biological hierarchy that is presently occurring. Neither the full dimensions and values of biological diversity nor the subtleties and ubiquity of its erosion are adequately represented and protected by the ESA. Moreover, the narrow view of diversity represented by the ESA easily leads to formulation of inadequate or inappropriate conservation strategies that detract from more sound ecosystem-level management.

The full dimensions of biodiversity necessary to understanding its nature and value are discussed first, including the value of animal populations vs. species and the distinction between local fauna and wildlife. Then, examples from the Southeastern Coastal Plain are presented that illustrate the breadth of erosion of the biodiversity resource and show that its measurement and conservation can only be achieved by integrated, regional programs that address landscape-level issues. Evaluation of the effectiveness of ESA as a tool for preserving biodiversity to date leads to the conclusion that the number of conflicts between species-oriented management recommendations will increase and further confuse decision makers to the point that much more of the biodiversity resource will be lost in the process. A new, broader-based program that focuses on integrated, interagency, regional systems management for biodiversity conservation is called for.

BIOLOGICAL DIVERSITY IS MORE THAN GENETIC OR SPECIES DIVERSITY

Biological diversity is "the variety and variability among living organisms and the ecological complexes in which they occur."[1] Although genetic diversity of organisms is propagated at the molecular level, phenotypic diversity, which humans identify with, is critically molded in a nonhereditary environmental context. It is the combination of both genetic and environmental elements working in concert that constitutes natural biological diversity. This notion has not yet become fully established within the conservation community, and there is a particular inclination on the part of some to equate genetic diversity with biological diversity. For example, one recent analysis[2] implied that even though the dusky seaside sparrow (*Ammospiza maritima nigrescens*) was formerly identified as a full species, its genetic constitution was not significantly divergent from other Atlantic coast subspecies, and little diversity was lost by its extinction. Such a reliance on genetic distinctions between organisms is as precarious as reliance on any other single characteristic. Species represent much greater entities of biological diversity than is measurable in their genes,[3] and most systematists would reject the notion that any single parameter (be it tail-length, pelage color, or allozyme frequency) is sufficient for differentiating species.

As an illustration, consider the classic case of four closely related species of eastern songbirds in the genus *Catharus* (hermit thrush, Swainson's thrush, gray-cheeked thrush, and veery). All are so visually similar as to befuddle analysts investigating museum skins alone. However, when species-specific song, nesting habitat, foraging habits, differences in both breeding and overwintering ranges, and analysis of morphometrics are conducted, there is little doubt that four distinct species exist.[4] Ecologists, naturalists, and even amateur bird watchers routinely evaluate such multiple dimensions in the process of describing biological diversity. However, current measures of genetic differentiation may not discriminate one taxon from the others.

A second example is from the ecosystem level and illustrates the necessity of the phenotypic/genotypic definition at a higher level. Numerous arguments have been marshalled for the conservation of old-growth forests, and some authorities have even argued that old growth is a nonrenewable resource.[5,6] Conversely, a consensus view from within the forestry profession is that old-growth forest can be regenerated by the creation of new stands planted with genetically improved nursery stock. Several important and untested notions underlie this assumption: (1) age is an adequate parameter for indexing senescence, (2) an old-aged, planted stand will develop into an old-growth forest, (3) a stand that is planted in a greatly altered landscape milieu today, and allowed to attain old age 500 years from now, is somehow equivalent to a forest that was regenerated by natural processes 500 years ago in a substantially different environment. Even if a stand were equal to a forest, and the genome planted now was the same as that removed, there is little prospect that the forces shaping a stand planted in a cutover landscape will be similar to the forces that shaped presently existing old-growth forests.

Both of the above examples illustrate a bothersome inclination to underrepresent the diversity of living, complex phenotypes by some narrow, albeit quantitative, measure of genotype. By analogy, these narrow definitions of biodiversity seem equivalent to accepting a collection of airlifted Kalahari Bushman fetuses for nurture in an Atlanta medical ward as being representative of the entire existing ethnic group.

Reductionism of this sort would suggest that a Savannah River Site (SRS) resource manager who wishes to increase the population of obligate cavity-nesting wood ducks should happily trade a tract of old-growth cypress (*Taxodium distichum*) containing hundreds of duck nesting cavities for a truck load of cypress seeds or seedlings that may or may not ever contain any nesting cavities when and if they mature. Biodiversity does indeed hinge on a genetic underpinning, but the full gamut of diversity to be protected is by no means captured by the genome or any other single level of hierarchy, including the species or the ecosystem. Biological diversity must be conserved at all scales ranging from the structure of the gene and pigments produced to the multicolored autumn panoramas that are created by gradients and disjunctures of these genes and pigments tempered by cold air currents passing through the Great Smoky Mountain landscape.

Finally, perhaps the most important, but the least often defended, component of biodiversity is the interconnectedness and dependence of different taxa upon each other and upon the ecosystem. This concept is central to understanding the economic and aesthetic value of almost any level of biodiversity. Attempts to reintroduce previously extirpated species commonly rely on degraded and incomplete environments, are expensive, and have low success rates under the best of circumstances.[7] Indeed, the very nature of a single species approach to conservation minimizes the importance of the interactive community *a priori*; yet it is the preservation of ecological interactions and interconnectedness that is perhaps the most valuable attribute of biodiversity.

The ESA has successfully focused public and professional attention on the erosion of biological diversity for the 20 years of its existence. Yet, precisely because of its success at riveting attention on species and their endangerment is the reason for moving now to broader based programs that will attract the public and financial support necessary to conserve biological diversity at all scales. The species focus of the ESA should and must be continued and amplified, just as every health care plan needs an intensive care unit. However, new programs based on stronger, more accurate definitions of biological diversity, its values, and the consequences of its loss must now be forged.

The work that wildlife does and the primary arguments for conserving wildlife do not hinge on the species. Even though increasingly strong ethical and moral arguments for conservation are being made[8,9] (these are ultimately best for developing a long-term preservation ethic), it seems that changes in human behavior must be built on more than aesthetic arguments. The primary values that humans derive from wildlife hinge on the day-to-day work that large, healthy populations and communities of wildlife do;[10] therefore, greater emphasis should be attached to this level of hierarchy when communicating the values of biological diversity to the general public. Just as the ecosystem services produced by a forest do not emanate from conservation of a relict woodlot, most of these values neither emanate from, nor are they perpetuated by, conservation of rare species.

The pharmaceutical industry that is undergirded by combinations of naturally occurring medicinal drugs may appear frustratingly cavalier in discounting the species within which those biochemicals occur. In most cases, a laboratory synthate is substituted for the natural product as soon as its biochemical structure is understood (e.g., acetylsalisylic acid substituted for salisylic acid). Thus, while biochemists do not marvel at the specific organism within which the product is found, there is great reason for them to celebrate and conserve the ecological processes of competition, predation, and herbivory that led to the product's evolution. In other words, it is the work of species interactions over millennia that has resulted in useful products, and it is the continued existence of these processes in natural systems that maintains their long-term utility.[11] Efforts of conservationists must be directed toward maintenance of the ecological processes that led to natural selection of these valuable species and products, not the conservation of static artifacts that can be collected and weighed.

The bureaus and businesses that tally tourist receipts on the basis of the fall colors created by maturing maple leaves (*Acer* spp.) in the Smoky or Adirondack Mountains should not be misled into thinking that theirs is a species diversity or species conservation issue. Barring total catastrophe, maple and other brightly colored leaves of foliar trees will exist in the future, but in the face of rapid, human-induced climatic changes, there is some doubt as to where or in what month humans would view them. Again, the aesthetic values that derive from this phenomenon have little to do with conservation of the species, but very much to do with conservation of the functioning environmental system within which the maples evolved and within which they occur.

The large numbers of people who live in coastal zones and have their livelihood protected from storm surges by coral reefs or coquina rocks (*Donax variabilis*) can be made aware of the value of reefs and shoals in their lives. However, a logical or compelling connection may not be obvious between the conservation of coral or coquina species in small, distant parks and the survival of people in coastal regions. Although it may be true that the rise in sea level is occurring too fast for the continued existence of living corals, no one has demanded conservation or captive breeding of coral species as a solution. Invoking single species conservation would be only remotely relevant to conserving the valuable functions of corals.

Similarly, millions of insects are necessary to affect biological control of crop pests and disease and to pollinate the hundreds of agricultural crops humans need for food and fiber. Man's dependence on crops, and thus on wildlife, is not significantly addressed by species conservation, but is critically dependent on the perpetuation of very large populations. To imply that simply saving an endangered species from extinction is important for utilitarian reasons is generally not true and quite probably misleading. Even in the case of undesirable pests and diseases, concern should not be with exterminating the species but with managing their population sizes.

Both geneticists and demographers concur that by the time populations have been reduced to a small number of individuals, much valuable genetic information has already been lost,[11-14] and these populations now face a suite of extinction forces not operative on large, viable populations.[14,15]

No matter whether it is the storehouse of genetic information, the recreational days of sport hunting provided, or the ecosystem services rendered that are of value, the conservation of large populations of wildlife occurring in their natural environment is what is of primary concern to humans and must be the primary focus for conservationists. Rather than withdrawing to the smallest defensible fragment of diversity, the endangered species, research and conservation efforts must be aggressively expanded to address the erosion of populations, ecological communities, and the quality of native regional environmental support systems.

FAUNA VS. WILDLIFE

Historian Alfred Crosby commented on the consequences of European discovery of the New World: "The long-range biological effects of the Columbian exchange are not encouraging. If one values all forms of life and not just the life of one's own species, then one must be concerned with the genetic pool. . . . We, all of the life on this planet, are the less for Columbus, and the impoverishment will increase."[16] The term "impoverishment" does not refer simply to the loss of species, such as the ivory-billed woodpecker (*Campephilus principalis*) or Carolina parakeet (*Conuropsis carolinensis*), but equally to the loss of faunal and floral identity. It refers in part to the homogenization of a formerly "hemispheric" world into an increasingly "homospheric" one, more specifically, the loss of faunal distinctiveness.[17-19] The full scope and magnitude of the erosion of biodiversity in the world today cannot be understood without reference to this homogenization process at all levels of biological hierarchy.

Unlike the term "fauna," which spans perhaps 200 years, coinage of the word "wildlife" is recent. All major treatises (e.g., Hornaday's *Wild Life Conservation in Theory and Practice*[20]) published early in this century distinguished between the terms, and even Leopold's text entitled *Game Management*[21] still used the term "wild life" rather than the single word "wildlife." Not until the charter of The Wildlife Society in 1937 did the single word gain official recognition.[22,23]

Current use of the term "wildlife" implies all free-ranging wild species and even some domestic species that have returned to wild habits (e.g., feral hogs). The meaning does not discriminate between endemic and exotic, wild or feral, pests or game, or for that matter, the biological phenomenon of rarity and the legal class of endangered. By losing sight of the concept of fauna, scientists have lost sight of an important level of biological diversity, the identifiably distinct regional floras and faunas. Widespread substitution of the term wildlife for fauna has not only facilitated loss of an important biological concept but also the erosion of biological diversity at one of the most critical levels, that of regional biological distinctiveness and regional identity.[23-28] Increasing the species richness of the southeastern United

States by 50 alien vertebrates that are already common elsewhere is not an acceptable trade for the loss of five endemics, such as the ivory-billed woodpecker, Carolina parakeet, red wolf, Florida panther, and Bachman's warbler.

FAUNAL COLLAPSE IN THE SOUTHEAST

The southeastern United States can serve as an example of the mechanisms and effects of the loss of fauna, both because this region began with a very rich and diverse fauna and because the history and processes have been relatively well studied there. At any given time, the fauna of a region consists of a dynamic balance between the number and type of species invading and those becoming locally extinct. Although this faunal diversity is dependent upon the type of habitats available, it is particularly sensitive to the areal extent and connectivity of indigenous habitats and how human-dominated ecosystems are introduced into the landscape. As acreage of native habitat decreases and becomes more fragmented, the fauna ultimately "relaxes" to some subset of the original complement.[30,31] Although the original formulation referred only to species loss, Harris[29] has extended the concept of faunal relaxation to include a number of other processes. To date, at least eight different measures of erosion of faunal diversity have been identified as operative in the Southeast.

Precipitous reduction of population numbers is the initial stage and most fundamental process leading to the loss of regional faunal character and distinctiveness. For example, the Everglades region of southern Florida now supports <5% of the wading birds that occurred in the system 100 years ago (Figure 1). Besides the loss of the birds themselves, this change has led to dramatic reductions in the amount of fish and aquatic invertebrate food consumed (from 21,000 to 3300 metric tons wet weight per year) and 100-fold decrease in the amount of nutrients imported to colony sites as feces (Figure 2). These differences are likely to have had large-scale effects on nutrient cycling within the relatively oligotrophic ecosystem[32,33] and on the demographics of prey populations.[34]

Inbreeding depression and the loss of diversity from within the genome results when small populations are forced to breed among themselves for extended periods of time. For example, of all adult male Florida panthers that have been investigated, about 95% of the spermatozoa are found to be congenitally malformed.[35] This is believed to be a direct result of long-term inbreeding among the individuals isolated in southwest Florida. Many other consequences of inbreeding depression have been documented, especially in captive populations,[36,37] and at least four correlates of fitness seem to be directly related to the erosion of genetic diversity.[38]

Loss of large, wide-ranging generalist species follows directly from reductions in total habitat area, its fragmentation into disjunct patches, and the amplified levels of mortality associated with moving from fragment to fragment in a human-dominated landscape. Rare, under the best of circumstances, large carnivores and many large territorial omnivores and herbivores are especially impacted by habitat

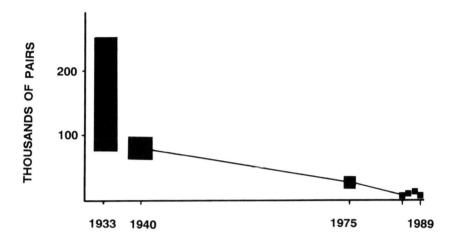

Figure 1. Estimated number of breeding pairs of wading birds in the Everglades ecosystem, from 1900 to the present, showing the range of estimates as upper and lower figures. The width of the bar indicates uncertainty in the estimate for any year or era. For all species except great egrets (*Casmerodius albus*), the decline has been between one and two orders of magnitude. Recent surveys show that only 7679 pairs of wading birds nested in the fresh and estuarine Everglades during 1989. (Data combined from References 55 to 62 and Everglades National Park Ranger Reports and National Audubon Society Warden Reports.)

fragmentation. Although systematic study of the demography of small populations has only recently occurred, Brussard and Gilpin[39] report that "the most important lesson concerning the biology of small populations is that stochastic processes play a critical role in their survival." Three types of stochasticity reduce survival probabilities: demographic uncertainty, such as sex ratio or number of offspring; environmental uncertainty, such as floods, droughts, or severe freezes; and genetic uncertainty, such as inbreeding and genetic drift. Vulnerability to local extinction appears to increase exponentially with decreasing population size.

Native southeastern species, such as the locally extinct red wolf (*Canis rufus*), the Florida panther (*Felis concolor coryi*), salt marsh mink (*Mustela vison lutensis*), manatees (*Trichechus manatus*), and wood storks (*Mycteria americana*), all manifest the consequences of hyperdispersed, rare populations with increased extinction probabilities. Because of their wide-ranging movements relative to the small size of parks and refuges, collisions with motor vehicles (automobiles and boats) have become the number one source of mortality for most of Florida's large, threatened, and endangered species (including the bald eagle *Haliaeetus leucocephalus*).[40]

Habitat specialists, such as the recently extirpated dusky seaside sparrow and Bachman's warbler (*Vermivora bachmanii*), or the closely related Swainson's warbler (*Limnothlypis swainsonii*), were historically rare for different reasons. Although this class of species may be locally abundant in high quality patches of habitat, these patches are usually small and/or widely dispersed. Because these

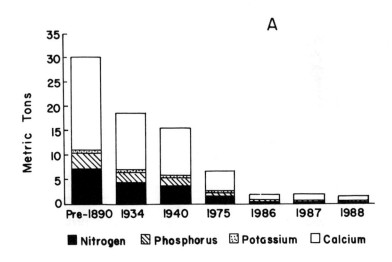
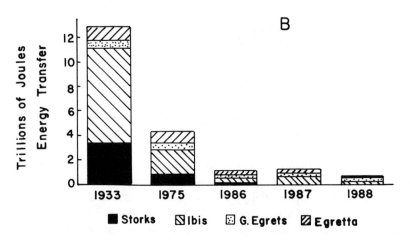

Figure 2. (A) Estimated dry weight of nutrients imported annually from wetland feeding sites to colony locations by colony-nesting wading birds in the Everglades ecosystem. (B) Energy in the form of nutrients flowing to colony sites, by species. Note both the overall decline of one order of magnitude (conservatively), as well as the drastic change in composition of species moving the nutrients. (Data derived from References 63 to 69 and Johnson and Bildstein, unpublished data.)

species are habitat specialists, they are highly sensitive to even limited degradation of habitat quality or encroachment by humans.

Increasing dominance by common, weedy species is facilitated by several of the processes mentioned above. Habitat fragmentation, the great increase in early successional or second-growth habitats, and the creation of new habitats amenable to these species facilitates the colonization and population expansion of opportunistic, weedy species of animals and plants that are adapted to human-dominated environments. Mammals, such as raccoon (*Procyon lotor*), gray fox (*Urocyon cinereoargenteus*), red fox (*Vulpes fulva*), and coyote (*Canis latrans*), continue to proliferate in the absence or tenuous presence of large native carnivores and continued urbanization of the landscape by humans. Similarly, a number of avian species, such as bronze-headed cowbirds (*Molothrus ater*), red-winged blackbirds (*Agelaius phoeniceus*), crows (*Corvus* spp.), blue jays (*Cyanocitta cristata*), and red-bellied woodpeckers (*Melanerpes carolinus*), prosper in human-dominated and agricultural landscapes and further jeopardize rare species, as explained below.

In addition to the direct alteration of native species frequency, **colonization by alien, exotic, and opportunistic species** may jeopardize the presently rare and endangered biota by altering primary ecological relations. Amplified competition for increasingly rare cavities exerted on the endangered red-cockaded woodpecker by increasingly common red-bellied woodpeckers and among colonially-nesting species for nest sites within colonies[41] are two well-known examples. Amplified ground nest predation by raccoons, feral dogs, pigs, and house cats becomes a critical factor in the continued existence of species, such as endangered marine turtles, gopher tortoise, colonially nesting wading and water birds, and neotropical migrant birds that nest on the ground. This is of particular concern in the Southeast because of the great abundance of endemic reptile and amphibian species, almost all of which nest on or slightly beneath the ground surface. Amplified levels of parasitism are manifest by the invasion and/or proliferation of obligate nest parasites, such as brown-headed cowbirds from the north and shiny cowbirds (*Molothrus bonariensis*) from the Caribbean. Mitigation of cowbird parasitism is now essential to the continued existence of Kirtland's warbler (*Dendroica kirtlandii*) elsewhere in the United States, and nest parasitism may well have been a factor in the extinction of Bachman's warbler.[42] Similarly, the exotic and fatal heartworm (*Dirofilaria immitus*) that is maintained in the southeastern environment by domestic dogs has already infected the red wolves that were only recently reintroduced into the Alligator River National Wildlife Refuge in North Carolina.[43]

Amplified levels of disease, such as rabies, distemper, and Lyme disease, are either exacerbated by, or result directly from, the invasion and unnaturally high population levels of opportunistic species, such as raccoons, foxes, and feral domestic dogs and cats. Not only do these diseases negatively impact numerous endangered species, but they also severely erode public support for faunal conservation in general. Endemic levels of feline panleukopenia is maintained in the environment by the collective population of wild and domestic cats, but only jeopardizes the Florida panther because of its low numbers and already tenuous survivorship.[35]

Regional faunal identity is eroded by all of the above processes. Species endangerment and loss is but one measure of the complex phenomenon of biodiversity erosion. Even though keeping an endangered species from extinction is most laudable, it does little to assure that either a functioning community or a suite of ecological relations similar to what existed in the native fauna is maintained (Figure 3). Actions directed specifically and solely at the conservation of species may be critical first steps, but these programs do not generally address either the root causes or the consequences of the biodiversity crisis. Indeed, aggressive efforts aimed at the conservation of one endangered species may even conflict with the conservation of another, especially in degraded habitats.

EXISTING CONSERVATION STRUCTURES ARE INADEQUATE

Florida and the Southeast have served as an incubator for North American conservation programming. Audubon's paintings of southeastern bird life and the 1905 murder of an Audubon warden in the Everglades focused international attention on the decimation of southeastern plume birds and the need for national conservation programming. Pelican Island on Florida's east coast was established as the first national wildlife refuge in the United States, and Ocala National Forest was the first designated national forest in the eastern United States. Unlike most early national parks that were established for their scenic grandeur, Everglades National Park was perhaps the first to be established primarily for wildlife and ecosystem conservation. More than 2.5 million acres of conservation land has been acquired in Florida during the last 25 years. Thirty-one species of vertebrates are federally listed as endangered; 11 are listed as threatened, and 65 are under review for federal listing.[44] With this level of programming, it seems reasonable to assess effectiveness and inquire whether amendments are required. Three examples illustrate that further modification and development of programming are urgently needed.

Florida's human population growth (4%/year) ranks with the fastest in the world, and habitat for native fauna is understandably decreasing at a rapid rate. Neither conservation programming at the land use and habitat preservation level nor at the species level are functioning to stem the loss.

For example, forest habitat acreage in Florida is being lost at a rate of 1%/year, over twice the rate of loss of forest in Brazil (Table 1). Bottomland hardwood forests and the old-growth longleaf pine (*Pinus palustris*) ecosystem type are disappearing even more rapidly, and several of the endemic species continue their historic decline. One endemic, the red-cockaded woodpecker (*Picoides borealis*), is federally listed as endangered, and the ESA has been invoked on several occasions to avoid effects of development. However, neither the ESA nor habitat acquisition programs have been effective at reversing the long-term decline.

The case of large mammals is equally challenging. Until 200 years ago, 11 native mammal species larger than 5 kg in size existed in Florida. Of these, the monk seal (*Monachus nonachus*) has been globally exterminated; the red wolf and bison (*Bison bison*) have been locally extirpated; the manatee, key deer (*Odocoilus virginiana clavium*), and Florida panther are federally listed as endangered; and the black bear, bobcat (*Lynx refus*), and otter (*Lutra canadensis*) are either listed by the state or by the Convention on Trade in Endangered Species (CITES). Road kill mortality is a very serious threat to wildlife in the Southeast, and collision by motor vehicles represents the number one known source of mortality for the following large threatened or endangered species: manatee, panther, key deer, black bear,

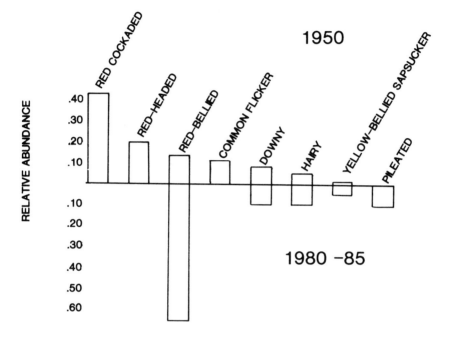

Figure 3. Species frequency distributions can be used to measure changes in faunal communities. The above data reflect changes in the relative abundance of woodpecker species in the pinelands of north-central Florida. Data for 1950 are drawn from Dennis;[70] data for the 1980s represent averages of pineland counts from several unpublished studies conducted in the Department of Wildlife and Range Sciences, University of Florida.

American crocodile (*Crocodylus acutus*), and bald eagle (*Haliaeetus leucocephalus*). It is questionable how the ESA can be invoked to prevent accidental mortality such as is occurring on Florida's highways and waterways.

As noted above, Everglades National Park was established specifically to facilitate the conservation of North America's unique subtropical wildlife. Since creation of the park, a total of 3 million acres of contiguous public land have been purchased to forestall development of surrounding Everglades habitat. The combination of parks, wildlife refuges, national preserves, and water management districts committed to conservation of Everglades water and wildlife now totals over 60% of the land area south of Lake Okeechobee. Still, not only are wading bird populations declining, but species are being lost from the park (Table 2). An even more aggressive integrated regional systems approach, such as an expanded and toughened regional biosphere reserve, is required.

Table 1. Comparative Forest Statistics for Several States in the Southeastern Coastal Plain and Several Latin American Countries[53,54]

State or Country	Forest Area Remaining	% Annual Loss
North Carolina	73	0.7
Georgia	90	0.4
Florida	60	1.0
Texas	48	0.3
Bolivia	300	0.2
Brazil	3575	0.4
Mexico	400	1.3
Panama	41	0.7
Peru	700	0.4
All Latin America (weighted x)		0.9

Endangered Species Conflicts: Alligators, Snail Kites or Wood Storks

The ESA does not provide adequate protection for many species, and it may also routinely generate management conflicts among endangered species, as well as between conservation of endangered species and ecosystem management. This results from the need to enhance conditions for a particular species in critical condition, even though such enhancements may not benefit other species. Conflicts between alligators (*Alligator mississippiensis*), snail kites (*Rostrhamus sociabilis plumbeus*), and wood storks in the Everglades ecosystem serve to illustrate the point.

During the late 1960s and early 1970s, the alligator constituted a premier endangered species that commanded both public and agency attention. Snail kites were already renowned for their diet specialization on apple snails (*Pomacea paludosa*). However, it was learned that young alligators also preyed heavily on the snails. As great emphasis was placed on increasing abundance of young alligators, predation pressure on *Pomacea* snails was inadvertently increased, and the snail kite food base was inadvertently affected. Because nesting success of snail kites is related to food abundance[45] and two predatory endangered species relied on the same prey species, balancing one endangered species against the another loomed as a problem.

Only modestly abundant under the best of conditions, the current population of snail kites undergoes population crashes in conjunction with drought-induced die-offs of the apple snail. No snail kites (the Everglades kite by former terminology) presently breed within Everglades National Park. Long periods of continuous inundation, lack of drydowns, and drought refugia seem critical for the continued existence of snail kites in North America.[46] As the Everglades system is increasingly dominated by man, the timing, nesting numbers, and nesting success of wood storks is increasingly dependent upon the extent and rapidity of surface water drydowns.

Table 2. List of Native Species of Birds and Mammals that No Longer Occur in Everglades National Park (in consultation with W. B. Robertson)

Native AmerIndian	Eastern bluebird
Monk seal	American kestrel
Red wolf	Red-cockaded woodpecker
Passenger pigeon	White-breasted nuthatch
Carolina parakeet	Brown-headed nuthatch
Ivory-billed woodpecker	Summer tanager

Only the most severe drydowns lead to successful nesting.[34,47,71] Clearly, the long period of inundation required by the snail kite is in direct conflict with the annual and severe drydowns currently needed to support wood stork reproduction. Undue emphasis on either species, as implied by a literal interpretation of the ESA, would be folly. A water regime that benefits the ecosystem as a whole must be considered.

In a world of perfect knowledge and nondegraded habitat, the ESA should be able to handle such conflicts, because both species existed in the same region prior to European colonization. However, the present lack of drought refugia for kites, lack of large numbers that would lend population resilience, and the general reduction in predictability and abundance of prey items for both species have accentuated the conflicting demands on ecosystem characteristics. The real problem is that the ecosystem is degraded, not that either species is somewhat threatened by persecution. Only by restructuring a regional Everglades management system that is large enough to accommodate different hydroperiod management regimes (where some pools are rapidly drawn down while others remain flooded) can the conflicting needs of the two endangered species be met. In any case, the ESA is not only unable to deal with the present degraded situation, it may even be hindering efforts to restore the ecosystem since advocates for either species may cite it as grounds for single species management.

The Health Care Analogy

The endangered species approach can be likened to an intensive care unit for fauna. It is never called to action until a species has been declared in critical condition, and attempts to improve the status of the species usually only proceed after severe habitat, population, and genetic damage have been done. The healing process for each patient (species) is expensive and often the patient must live on contrived life-support systems following intensive care. In this era of increasing degradation of habitats and species, the intensive care units will be routinely swamped, have a long waiting list, and turn away patients (species) that are not on the verge of extinction. To complete the analogy, any reasonable society would retain the intensive care unit for emergencies, but would implement an aggressive preventive maintenance program.

This points up the bare bones of the biodiversity problem: the erosion of biodiversity can only be abated by a fundamental change in public attitudes and a major increase in the scale of resource conservation thinking. Quite frankly, the ethic is questionable by which a single species in critical condition (e.g., the spotted owl, *Strix occidentalis*, or red-cockaded woodpecker) must be "used" to rationalize the conservation of old-growth forests. Old-growth forests must be conserved for many scientific, moral, and aesthetic reasons, only one of which involves an endangered species. The single species approach is also likely to backfire for several reasons. First, Everglades-type species conflicts and the generally poor success at saving species such as the dusky seaside sparrow and Bachman's warbler will begin to erode public confidence in the approach. Second, the root problems of ecosystem degradation that are highlighted by the biodiversity crisis can be far more effectively and economically addressed if they are confronted directly. The only effective approach to the biodiversity crisis will depend upon recognition of its full scope and root causes, not simply band-aid prescriptions for the most obvious symptoms.

NECESSARY NEW PROGRAMMING

Biological diversity occurs at many levels of scale (ranging from within-species genetic diversity through and including the identity of the faunal region), and erosion of biodiversity is occurring at all scales. The objective of biodiversity conservation should be the maintenance of large, functional populations and the original character of the presettlement regional fauna, not simply the survival of species. Existing institutional constructs do not seem adequate to meet the challenge of conserving the regional fauna. Regional scale, interagency, habitat-focused programming is necessary to achieve the task of biodiversity conservation. As suggested by Odum,[48] only a hierarchical approach ("top-down" or "outside-to-inside"), where internal systems behavior is understood to be governed by landscape-level forces and constraints, can address the crisis situation that presently exists.

For the most part, only species with small body size, specialized habits, and/or local distributions have sufficiently identifiable habitat requirements such that critical habitat can be designated. Species with large body size have wide home ranges and tend to be generalists. Thus, it is little wonder that critical habitat could be designated for the dusky seaside sparrow but cannot be designated for the Florida panther.

Conservation efforts during this decade illustrate that the maintenance of minimum viable populations of most organisms will require management at the regional level (i.e., often two or more states).[49,50] No single park or preserve in North America appears big enough in and of itself to maintain its original complement of species.[51] The most promising strategy hinges on strategically chosen parks, preserves, or refuges that serve as unmanaged biodiversity islands which are, in turn, surrounded by public domain lands that are managed specifically to control regional forces and constraints (i.e., the inputs). Lands, such as national forests, water management districts, and those of the Bureau of Land Management and the Department of Defense, are ideal.

Because habitat specialists generally have small home ranges and local distributions that can be designated as "critical habitat," core areas of the biodiversity complexes are ideally chosen to conserve viable populations of these species. The large, charismatic species, such as the Florida panther, red wolf, and black bear, demand extensive areas, but are highly tolerant to most forms of low-intensity land use. Thus, the characteristics of this strategy consist of identification of three or more strategic focal areas that serve as critical habitat for specialist species. These biodiversity islands are then surrounded by a managed regional landscape so that inputs to the biodiversity island are known, monitored, and controlled if necessary. At the same time that the park or preserve is "buffered" from external forces, such as exotic species, pollution, or highly unnatural hydroperiod, the humans who occupy the regional landscape are "buffered" from the depredations of high populations of native species (e.g., crop- or stock-raiding species). Leases, easements, and quid-pro-quo covenants must be signed to encourage each cooperating party to participate in their authorized and responsible capacity for the maintenance of regional ecosystem function, including the viability of such wide-ranging species as panthers, bears, wood storks, and red wolves.

In most cases, these **multiple use modules**[27,28,52] will not require extravagant purchases of land, but will require key acquisitions and easements that ensure connectivity between existing preserves and refuges. Most critically, this plan requires an interagency and/or interstate management covenant that designates the specific responsibilities of several different agencies ranging from state agencies to the USDI, USDA, USDOD, Indian reservations, and even the private sector. About a dozen such areas have been investigated in the southeastern coastal plain, and, in many cases, the amount of land necessary to aggregate and solidify a million-acre complex of public domain land is as little as 10,000 acres.

Various sections of the ESA could play key roles in implementation of such a biodiversity conservation strategy. The critical habitat clause (section 7) could be invoked to provide focus on key regional sites. The cooperative agreements, and interagency consultation clauses (sections 6, 7) could be used to stimulate and implement interagency covenants. The land and water conservation fund constitutes a funding source for the purchase of critical corridors, linkages, and buffers to ensure connectivity and cohesiveness of the land complexes.

Two central questions that must be answered when designing and managing such regional systems are what were conditions like prior to degradation, and how will the achievement of a functioning and integrated ecosystem be known? The first question is difficult because few reserves have sufficient historical data upon which to base a regional ecosystem model. In some cases, it is difficult to say with certainty which species occurred in the area, and rarely is their relative abundance known.

The second question hinges on adaptive environmental management. In the absence of historically derived models, reconstructions and alterations to management plans must proceed as an iterative, closely monitored process. Research on the ecology, demographics, and behavior of the system and its keystone species will, of necessity, be simultaneous with adaptive management of the regional system.

ACKNOWLEDGMENTS

The work reported here is supported by contracts and grants from the U.S. National Park Service and the U.S. Man and The Biosphere Program to L. D. Harris, and from the U.S. Army Corps of Engineers to Dr. Michael Collopy.

REFERENCES

1. Office of Technology Assessment. "Technologies to Maintain Biological Diversity," (Washington, D.C.: U.S. Government Printing Office, OTQ-F-330, 1987).
2. Avice, J., and W. Nelson. "Molecular Genetic Relationships of the Extinct Dusky Seaside Sparrow," *Science* 243:646-648 (1989).
3. Lewontin, R., S. Rose, and L. Kamin. *Not in Our Genes* (New York: Pantheon Books, 1984).
4. Mayr, E. *Animal Species and Evolution* (Cambridge, MA: Harvard University Press, 1966).
5. Franklin, J., and T. Spies. "Characteristics of Old-Growth Douglas-Fir Forests," in *New Forests for a Changing World, Proceedings of the Society of American Foresters National Convention* (Bethesda, MD: Society of American Foresters, 1983), pp. 328-334.
6. Thomas, J., R. Lancia, R. Mannan, L. Ruggiero, and J. Schoen. "Management and Conservation of Old-growth Forests in the United States," *Wildl. Soc. Bull.* 16(3):252-262 (1988).
7. Griffith, B., J. Scott, J. Carpenter, and C. Reed. "Translocation as a Species Conservation Tool: Status and Strategy," *Science* 245:477-480 (1989).
8. Norton, B., Ed. *The Preservation of Species* (Princeton, NJ: Princeton University Press, 1986).
9. Rolston, H. "Duties to Endangered Species," *BioScience* 35:718-726 (1985).
10. Ehrlich, P. "The Loss of Diversity, Causes and Consequences," in *Biodiversity*, E. Wilson, Ed. (Washington, D.C.: National Academy Press, 1988), pp. 21-27.
11. Oldfield, M. *The Value of Conserving Genetic Resources* (Sunderland, MA: Sinauer Associates, Inc., 1989).
12. Frankel, O., and M. Soule. *Conservation and Evolution* (New York: Cambridge University Press, 1981).
13. Schonewald-Cox, C., S. Chambers, B. MacBryde, and W. Thomas. *Genetics and Conservation* (Menlo Park, CA: Benjamin Cummings, 1983).
14. Soule, M., Ed. *Viable Populations for Conservation* (New York: Cambridge University Press, 1987).
15. Seale, U., E. Thorne, M. Bogan, and S. Anderson, Eds. *Conservation Biology and the Black-Footed Ferret* (New Haven, CT: Yale University Press, 1989).
16. Crosby, A. *The Columbian Exchange, Biological and Cultural Consequences of 1492* (Westport, CT: Greenwood Press, 1972).
17. Allen, J. "On the Mammals and Winter Birds of East Florida with an Examination of Certain Assumed Specific Characters in Birds, and a Sketch of the Bird Faunae of Eastern North America," *Bull. Mus. Comp. Zool.* 3:161-450 (1871).
18. Merriam, C. "The Crop Zones and Life Zones of North America," USDA Bureau of Biological Survey Bulletin Number 10 (1898).

19. Dice, L. *The Biotic Provinces of North America* (Ann Arbor, MI: University of Michigan Press, 1943).
20. Hornaday, W. *Wild Life Conservation in Theory and Practice* (New Haven, CT: Yale University Press, 1914).
21. Leopold, A. *Game Management* (New York: Charles Scribner's Sons, 1933).
22. Hunter, M. *Wildlife, Forests, and Forestry: Principles of Managing Forests for Biological Diversity* (Englewood Cliffs, NJ: Prentice Hall, in press).
23. Harris, L. "The Faunal Significance of Fragmentation of Southeastern Bottomland Forests," in *Forested Wetlands of the South*, D. Hook, Ed. (Asheville, NC: USDA Forest Service, 1988), GTR SE-50.
24. Noss, R. "A Regional Landscape Approach to Maintain Diversity," *BioScience* 33:700-706 (1983).
25. Swanson, F., and F. Knopf. "In Search of a Diversity Ethic for Wildlife Management," *Trans. N. Am. Wildl. Nat. Res. Conf.* 47:421-431 (1982).
26. Faaborg, J. "Potential Uses and Abuses of Diversity Concepts in Wildlife Management," *Trans. Miss. Acad. Sci.* 14:41-49 (1980).
27. Harris, L. *The Fragmented Forest: Application of Island Biogeography Principles to Preservation of Biotic Diversity* (Chicago: University of Chicago Press, 1984).
28. Noss, R., and L. Harris. "Nodes, Networks, and MUMs: Preserving Diversity at all Scales," *Environ. Manage.* 10:299-309 (1986).
29. Harris, L. "The Nature of Cumulative Impacts on Biotic Diversity of Wetland Vertebrates," *Environ. Manage.* 12:675-693 (1988).
30. Diamond, J. "Biogeographic Kinetics: Estimation of Relaxation Times for Avifaunas of Southwest Pacific Islands," *Proc. Natl. Acad. Sci. U.S.A.* 69:3199-3202 (1972).
31. Terborgh, J. "Preservation of Natural Diversity: The Problem of Extinction Prone Species," *BioScience* 24:715-722 (1974).
32. Stinner, D. H. "Colonial Wading Birds and Nutrient Cycling in the Okefenokee Swamp," PhD Thesis, University of Georgia, Athens (1983).
33. Bildstein, K. L., E. Blood, and P. C. Frederick. "The Relative Importance of Biotic and Abiotic Vectors in Nutrient Processing in a South Carolina, U.S.A. Estuarine Ecosystem," *Estuarine, Coastal and Shelf Science* (in press).
34. Kushlan, J. A. "Responses of Wading Birds to Seasonally Fluctuating Water Levels: Strategies and Their Limits," *Col. Waterbirds* 9:155-162 (1986).
35. U.S. Fish and Wildlife Service. "Florida Panther Recovery Plan," prepared for the U.S. Fish and Wildlife Service by the Florida Panther Interagency Committee, Atlanta (1987).
36. Ralls, K., and J. Ballou. "Genetic Diversity in California Sea Otters: Theoretical Considerations and Management Implications," *Biol. Conserv.* 25:209-232 (1983).
37. Ralls, K., J. Ballou, and A. Templeton. "Estimates of Lethal Equivalents and the Cost of Inbreeding in Mammals," *Conserv. Biol.* 2:185-193 (1988).
38. Quattro, J., and R. Vrijenhoek. "Fitness Differences Among Remnant Populations of the Endangered Sonoran Top Minnow," *Science* 245:976-978 (1989).
39. Brussard, P., and M. Gilpin. "Demographic and Genetic Problems of Small Populations," in *Conservation Biology and the Black-Footed Ferret*, U. Seal, E. Thorne, M. Bogan, and S. Anderson, Eds. (New Haven, CT: Yale University Press, 1989).
40. Harris, L. "New Initiatives for Wildlife Conservation, the Need for Movement Corridors," in *In Defense of Wildlife: Preserving Communities and Corridors*, G. MacKintosh, Ed. (Washington, D.C.: Defenders of Wildlife, 1989), pp. 11-34.

41. Burger, J. "The Pattern and Mechanism of Nesting in Mixed-Species Heronries," in *Wading Birds*, Research Report 7, I. V. Sprunt, J. C. Ogden, and S. Winckler, Eds. (New York: National Audubon Society, 1978).
42. Hamel, P. *Bachman's Warbler: A Species in Peril* (Washington, D.C.: Smithsonian Institution Press, 1986).
43. Phillips, M., and J. Schneck. "Clinical Observations of Parasitism in Reintroduced and Captive Red Wolves," *J. Wildl. Dis.* (in press).
44. Wood, D. "Official Lists of Endangered and Potentially Endangered Fauna and Flora in Florida," Florida Game and Freshwater Fish Commission, Tallahassee (1988).
45. Beissinger, S. "Demography, Environmental Uncertainty, and the Evolution of Mate Desertion in the Snail Kite," *Ecology* 67:1445-1459 (1986).
46. Bennetts, R. E., M. W. Collopy, and S. R. Beissinger. "Nesting Ecology of Snail Kites in Water Conservation Area 3A," Department of Wildlife and Range Science, University of Florida, Gainesville, FL. Florida Cooperative Fish and Wildlife Research Unit, Technical Report 31 (1988).
47. Kushlan, J. K., J. C. Ogden, and J. Tilmant. "Relation of Water Level and Fish Availability to Wood Stork Reproduction in the Southern Everglades, Florida," U.S. Geological Survey Open File Report 75-434 (1975).
48. Odum, E. "Input Management of Production Systems," *Science* 243:177-182 (1989).
49. Salwasser, H. "Managing Ecosystems for Viable Populations of Vertebrates: A Focus for Biodiversity," in *Ecosystem Management for Parks and Wilderness*, J. Agee and D. Johnson, Eds. (Seattle: University of Washington Press, 1988), pp. 87-104.
50. Salwasser, H., C. Schonewald-Cox, and R. Baker. "The Role of Interagency Cooperation in Managing for Viable Populations," in *Viable Populations for Conservation*, M. Soule, Ed. (New York: Cambridge University Press, 1987), pp. 159-173.
51. Newmark, W. "A Land-Bridge Perspective on Mammalian Extinctions in Western North American Parks," *Nature* 325:430-432 (1987).
52. Harris, L. "An Island Archipelago Model for Maintaining Biotic Diversity in Old-Growth Forests," in *New Forests for a Changing World, Proceedings of the Society of American Foresters National Convention* (Bethesda, MD: Society of American Foresters, 1983), pp. 60-64.
53. Molofsky, J., C. Hall, and N. Myers. "A Comparison of Tropical Forest Surveys," Report Number TR032, DOE/NBB-0078, U. S. Department of Energy, Washington, D.C. (1986).
54. U.S.D.A. Forest Service. "The South's Fourth Forest: Alternatives for the Future," Forest Resource Report No. 24, Washington, D.C. (1988).
55. Robertson, W. B. and J. A. Kushlan. "The Southern Florida Avifauna," *Miami Geol. Soc. Mem.* 2:414-452 (1974).
56. Kushan, J. A. "Population Energetics of the American White Ibis," *Auk* 94:114-122 (1977).
57. Kushlan, J. A. and D. A. White. "Nesting Wading Bird Populations in Southern Florida," *Fla. Sci.* 40:65-72 (1977).
58. Kushlan, J. A., P. C. Frohring and D. Vorhees. "History and Status of Wading Birds in Everglades National Park," National Park Service Report, South Florida Research Center, Everglades National Park, Homestead (1984).
59. Ogden, J. C. "Recent Population Trends of Colonial Wading Birds on the Atlantic and Gulf Coastal Plains," in *Wading Birds*, A. Sprunt, IV, J. C. Ogden, and S. Winckler, Eds., National Audubon Society Research Report 7:135-153 (1978).

60. Kushlan, J. A. and P. C. Frohring. "The History of the Southern Florida Wood Stork Population," *Wilson Bull.* 98:368-386 (1988).
61. Ogden, J. C., D. A. McCrimmon, G. T. Bancroft and B. W. Patty. "Breeding Populations of the Wood Stork in the Southeastern United States," *Condor* 89:752-759 (1987).
62. Frederick, P. C. and M. W. Collopy. "Nesting Success of Five Ciconiiform Species in Relation to Water Conditions in the Florida Everglades," *Auk* 106:625-634 (1989).
63. Kahl, M. P. "Bioenergetics of Growth in Nesting Wood Storks," *Condor* 64:169-183 (1962).
64. Kahl, M. P. "Food Ecology of the Wood Stork (*Mycteria americana*) in Florida," *Ecol. Monogr.* 34:97-117 (1964).
65. Kushlan, J. A. "The Ecology of the White Ibis in Southern Florida, a Regional Study," PhD Thesis, University of Miami, Miami (1974)
66. Kushlan, J. A. "Growth Energetics of the White Ibis," *Condor* 79:31-36 (1977).
67. Wiens, J. A. and M. I. Dyer. "Assessing the Potential Impact of Granivorous Birds in Ecosystems," in *Granivorous Birds in Ecosystems: Their Evolution, Populations, Energetics, Adaptations, Impact and Control*, J. Pinowska and S. C. Kendeigh, Eds. (New York: Cambridge University Press, 1977).
68. Kendeigh, S. C., V. R. Dol'nick and V. M. Gaurilov. "Avian Energetics," in *Granivorous Birds in Ecosystems: Their Evolution, Populations, Energetics, Adaptations, Impact and Control*, J. Pinowska and S. C. Kendeigh, Eds. (New York: Cambridge University Press, 1977).
69. Mock, D. W., T. C. Lamey and B. J. Ploger. "Proximate and Ultimate Roles of Food Amount in Regulating Egret Sibling Aggression," *Ecology* 68:1760-1772 (1987).
70. Dennis, J. "A Comparative Study of Florida Woodpeckers in the Non-Breeding Season," MSc Thesis, University of Florida, Gainesville (1951).
71. Ogden, J. Personal communication, 1978.

CHAPTER 11

Endangered Species Protection — The Wood Stork Example

William D. McCort and Malcolm C. Coulter

INTRODUCTION

Extinction of species and the loss of biodiversity are accelerating and forebode major disruptions of the earth's ecosystem. The situation is so serious that even the survival of human civilization is at risk.[1] Unrestrained human exploitation of the earth's resources is causing the problem, and human decision making and action must solve it.

Endangered species protection is a vital attempt to prevent the extinction of species and to allow the recovery of species threatened with extinction. The process is often difficult and expensive. Government, industry, agriculture, environmental groups, academic institutions and virtually all consumers and taxpayers are involved in and affected by decisions regarding endangered species protection.

The development and implementation of environmental management plans must include consideration of endangered species issues and must provide a forum for discussion of endangered species protection. A specific case history of the wood stork (*Mycteria americana*) is presented here as an example of a successful solution to an endangered species protection problem.

STATEMENT OF THE PROBLEM

The problem is not that species become endangered and eventually extinct, since extinction is a natural process of evolution. Rather the problem is the rapidly increasing *rate* at which species are currently becoming endangered and extinct. The rapid increase in the world human population (Figure 1),[2-5] and the concurrent increase in the number of extinct species (Figure 2)[6] are not just a coincidence. Human population is increasing at an alarming rate, is degrading the earth's environment at an equally alarming rate, and is responsible for the increased rate of extinctions.[1,7]

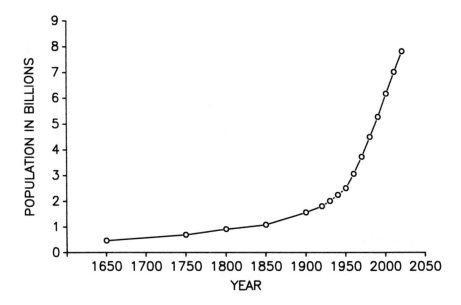

Figure 1. World human population and projections for the years 1650 to 2020.

The tropical rain forest is one of many threatened habitat types throughout the world. It is, however, a critically important example because of its high rate of destruction and because of its extraordinary species diversity. E. O. Wilson[7] has analyzed the loss of tropical rain forests and the associated loss of biodiversity as follows. There are 1.4 million species known worldwide. Most species have yet to be discovered and described; in actuality, there are probably 5 to 30 million species in existence. Tropical rain forests are astonishingly rich in the number of species that they harbor, most not described. Although tropical rain forests cover only 7% of the earth's land surface, they contain more than 50% of all species. Currently, tropical forests persist on only about 60% of all land area that according to bioclimatic data may have at one time supported such habitat.[7,8] Furthermore, tropical forests are being destroyed at a rate of 76,000 km²/year. At this rate, all tropical forests will be either clear-cut or seriously disturbed by the year 2135 AD, close to the date (2150) that the human population will reach 11 billion. This is a conservative estimate using the lower range of deforestation estimates, 76,000 to 92,000 km²/year, and assum-

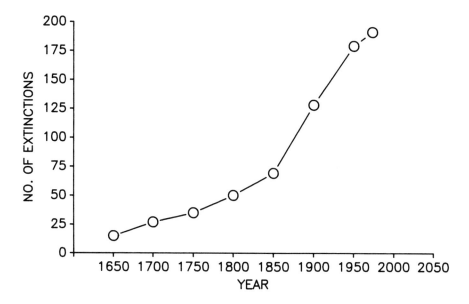

Figure 2. Number of extinct species from the years 1600 to 1973.

ing a constant rate of deforestation.[8] A rough approximation of the extinction rate, given an estimate of 5 million tropical forest species, is 17,500 species per year. This is about 1000 to 10,000 times faster than the rate of extinction before the influence of humans!

RATIONALE FOR PROTECTING ENDANGERED SPECIES

There are many good reasons for protecting endangered species. Animals and plants are important to humans for food, clothing, medicine, aesthetics, and many other human oriented needs. The majority of plants and animals have yet to be identified, and their potential importance to humans has not been explored. Protection of plants and animals has also been justified based upon respect and the recognition of their right to life.

Although important and valid, all of the above reasons for protecting endangered species pale in importance when compared to the need to preserve the earth's ecosystem upon which all life depends. Many highly visible and popular species have been recognized as endangered and worthy of protection. However, tens of thousands of seemingly minor or totally unknown species are being lost. The loss is undermining the stability of ecosystems upon which species, including humans, depend for survival.

Ehrlich[1] describes this loss in terms of "ecosystem services." For example, all plants and animals play a role in maintaining the mix of gases in the atmosphere. Changes in the gaseous composition of the atmosphere by elimination of populations and species (e.g., the massive deforestation of the tropics) could cause changes in the global climate and result in agricultural disaster. Such a climate change could result in the starvation and death of as many as 1 billion people before 2020. Liberal application of pesticides can exterminate natural predators of insect pests and lead to major pest epidemics. Extinction of subterranean organisms can disrupt soil fertility. Ehrlich's point is that plants and animals play roles in ecological systems that are essential to human civilization.

It is very difficult to predict the importance of any particular species since the roles and interrelationships of most species are not well understood. The extinction of a single species may not have a significant impact on the ecosystem. Similarly, the removal of one brick from a building may not matter. The continued loss of species, however, will lead to a cataclysmic collapse of the ecosystem, just as the continued removal of bricks will result in the collapse of a building.[9] This then, the need to maintain ecosystem stability, is the fundamental reason for protecting endangered species.

THE ENDANGERED SPECIES ACT

In 1973, Congress passed the Endangered Species Act (ESA). It resolved the shortcomings of earlier endangered species legislation, the Endangered Species Preservation Act of 1966 and the Endangered Species Conservation Act of 1969. The ESA requires that federal agencies protect all species of plants and animals facing possible extinction. Two categories of such species are specified as (a) endangered, which includes all species in danger of extinction throughout all or a significant part of their ranges, and (b) threatened, all species likely to become endangered within the foreseeable future throughout all or a significant portion of their ranges.[10] The act makes it a crime to violate endangered species regulations and supports the protection of critical habitats. After a species is listed the U. S. Fish and Wildlife Service (USFWS) is charged with preparing a species recovery plan.[10] Individual states have also adopted their own lists and plans for protection of endangered and threatened species, and the International Union for Conservation of Nature and Natural Resources maintains the Red Data Book listing species worldwide that are in danger of extinction.

The number of species officially listed as endangered or threatened is increasing, but at a fairly slow rate (Figures 3 and 4).[11] In the United States, there are currently listed (as of April 30, 1989) 407 endangered species (249 animals, 158 plants) and 128 threatened species (82 animals, 46 plants). The number of species currently listed in foreign countries as endangered is 516 (509 animals, 7 plants); the number listed as threatened is 60 (52 animals, 8 plants). There are currently 242 approved recovery plans.[12]

The slow rate at which new species are being added to the official listings does not accurately reflect the large number of species threatened with extinction. Less than two years after the ESA was passed, the USFWS had received petitions to list

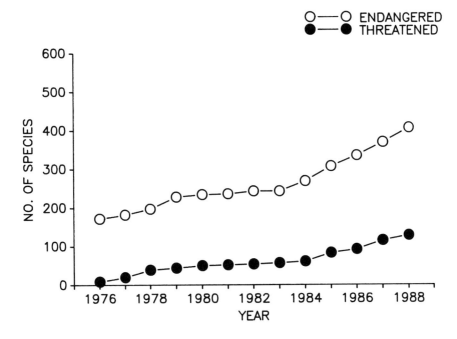

Figure 3. Number of species listed as endangered and threatened inside the United States from 1976 to 1988.

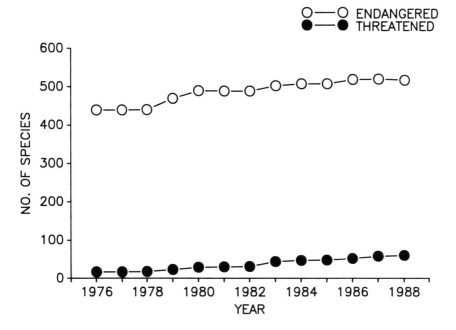

Figure 4. Number of species listed as endangered and threatened outside the United States from 1976 to 1988.

23,962 species in addition to the 144 listings that it had initiated.[13] As of 1988, the USFWS had about 3894 candidate species for listing. The process of adding new species to the list is complicated, slow, and can be very controversial. The first and most famous controversy, the snail darter versus the Tellico Dam, made it clear that protection of endangered species can be challenging, costly, and political.

THE WOOD STORK EXAMPLE

The wood stork, *Mycteria americana* (Figure 5) is a member of the stork family, Ciconiidae, which includes 17 to 19 species, depending upon which classification is used.[14] It is a large wading bird ranging from the southeastern United States to Paraguay and southern Brazil.[14] In the United States, the wood stork is the largest wading bird and the only breeding stork. There is no known interchange between the southeastern United States stork population and storks that nest in southern Mexico.[15] The USFWS classified the United States population of wood storks as endangered in 1984.[16]

The wood stork is not considered endangered throughout its range outside of the United States. However, agricultural development of wetlands in the Gulf of Mexico, the Venezuelan llanos, and the Brazilian pantanal, as well as hunting and egg collecting at colony sites throughout Latin America, threaten the species.[17] Luthin,[17] in a survey of the world's stork species, found that wood storks are regionally threatened as follows. Populations may be declining in Chiapas, Tabasco, and Campeche States of Mexico. The wood stork is considered endangered in Belize. All of the nesting colonies known to exist in Belize during the 1970s have disappeared due to hunting and disturbance. There is only one known stable colony of wood storks in Costa Rica. The status of the wood stork in other Central American countries is not known. Wood storks are common in Venezuela and the Brazilian pantanal, but are threatened in Sao Paulo and Rio Grande do Sul, Brazil.

Status in the United States

Wood storks in the United States have been reported to nest in all of the coastal states from Texas to South Carolina.[18] Documentation of wood storks nesting in southern Florida dates to 1880.[19] The number of breeding pairs of wood storks in the southeastern United States has decreased from an estimated 20,000 pairs in the 1930s, to about 10,000 pairs in the 1960s, and to 5000 to 6000 pairs in 1975-1976.[15]

Accurate estimates of numbers of wood storks are difficult to obtain. Censuses have been limited to colony sites where wood storks congregate for nesting. The number of nesting wood storks in a colony varies from year to year depending upon feeding and nesting conditions. Hence, the most dependable numbers are obtained during years when nesting and feeding conditions are best.[20] Historical records are often incomplete or inaccurate. After review of the historical records, Kushlan and Frohring[19] reported that they found no evidence that southern Florida has ever

Figure 5. Wood stork in flight carrying nesting material to nest in the Birdsville Colony, Georgia. Photograph by Walker Montgomery, University of Georgia.

supported more wood storks than the maximum estimates of 8000 pairs in 1912, 8535 pairs in 1960 and 9425 pairs in 1967.

Population estimates from the late 1950s to the present have been more reliable and confirm a decrease in the number of wood storks. Ogden and Nesbitt[18] reported a 41% decrease in the number of breeding pairs of wood storks from 10,060 pairs in 1960 to 5982 pairs in 1975. Kushlan and Frohring[19] reported a 75% decrease in the southern Florida population from 1967 to 1981-1982.

Ogden and colleagues[15] demonstrated two new trends since the late 1970s. The total number of nesting wood stork pairs stabilized and possibly increased to about 6000 pairs; also, nesting increased in the northern colonies but decreased in the southern colonies. Over 6000 pairs of storks nested in 1983 (6097) and 1984 (6447) and over 5000 in 1985 (5355). These totals are higher than those recorded in any year after 1977. However, the increases were recorded only for more northern colonies. Colonies in southern Florida, formerly the site for the largest number of nesting storks, declined 84% from 8835 pairs in 1960 to 1455 pairs in 1985. From 1958 through 1960, the center of the breeding range was south of Lake Okeechobee, Florida. The center moved 87 km north during 1976 through 1979 and another 45 km north between 1980 and 1985. The mean annual number of wood stork pairs nesting in the north increased from 2082 (1958 to 1960) to 2704 (1976 to 1979), and to 3547 (1980 to 1985). In Georgia and South Carolina, the most northerly colony sites, nesting pairs increased from 0 in 1960 to 605 in 1985.

Causes of Decline

In the United States, the most apparent reason for the decrease in the number of wood storks is reduced breeding success. Other causes, such as hunting, collection for feathers, lack of protection from human disturbance, and chemically related mortalities, are not supported in historical or recent records. Ogden and Patty[20] determined that the average number of young wood storks produced per pair of adults from 1977 to 1981 was 0.77 (range 0.45 to 1.24). Such a low annual reproductive rate would not be enough to maintain a stable population. The European white stork (*Ciconia ciconia*), for example, must produce 1.5 young per pair of adults in order to maintain a stable population. Using the European white stork as a model, Ogden and Patty calculated that, theoretically, the wood stork breeding population would decline about 5.3% annually. A mean annual rate of decline of 4.4% was recorded from 1975 to 1980 when the stork population decreased from 6000 pairs to 4800 pairs.

The major cause for the decline of the wood stork breeding success is the reduction in foraging habitat and food availability. As a tactile feeder, the wood stork depends on shallow wetland areas with high densities of fish. Most commonly, such wetland feeding areas are created by evaporation that reduces water levels and concentrates fish. Increased agricultural and urban land development and water management practices in southern Florida have altered foraging sites so that they no longer produce enough fish or have changed the water regime in wood stork foraging areas so that water depths are too deep to concentrate for feeding during the appropriate times. Ogden and Nesbitt[18] determined that the most important feeding habitats of storks (cypress domes and strands, wet prairies, scrub cypress, freshwater marshes and sloughs, and sawgrass marshes) have been reduced by 35% since 1900.

The availability of nesting habitat may be a secondary factor in the decreased breeding success of the wood stork.[15,20] An increase in the percentage of storks nesting in man-made impoundments and mangrove islands may indicate that cypress swamps are becoming less available for nesting.[18] Drainage and water management have resulted in cypress swamps being dry more often. Since wood storks nest in trees over water, the number of suitable cypress nesting sites has been decreasing.[18]

The Birdsville, Georgia Colony

The Birdsville colony is located in Big Dukes Pond, a 567-ha cypress forested Carolina bay[21] near Birdsville in the township of Millen, Jenkins County, GA. This colony is of particular importance and interest since it is the most northern and inland colony in the southeastern United States. The colony is likely a part of the shift of wood stork colonies north. Since most studies of wood storks have been in Florida, it is important to study the Birdsville colony in order to understand the ecology and availability of wood stork habitat in the north. The colony has received a great deal of attention since 1983 when wood storks from the colony were observed foraging on the Department of Energy's Savannah River Site near Aiken, SC. Since 1983, the University of Georgia's Savannah River Ecology Laboratory has conducted intense research on the Birdsville colony.[22-27]

Wood storks have arrived at the Birdsville colony from March to April each year since it was discovered in 1980.[28] The number of breeding pairs of wood storks, or the number of nests in the colony, remained fairly stable from 1980 to 1985 at 100, unknown, 60, 113, 100, and 108 nests per year, respectively. An unknown number of pairs abandoned their nests in 1981 because of dry conditions. The number of nests increased considerably to 160 in 1986 and to 193 in 1987. There were 101 active nests in 1988.[27]

During most years, reproductive success within the colony has been well over the minimum of 1.5 young per pair of adults necessary to maintain a stable population.[20] However, there were years of poor or no successful reproduction. The average number of fledglings per nest ranged from 0 to 2.19 from 1980 to 1988. Despite the large increase in the number of nesting pairs in 1986 to 1988, the total number of fledglings did not increase.[26,27]

Wood storks leaving the Birdsville colony to forage were routinely followed by airplane. The average distances that wood storks flew to foraging sites from the Birdsville colony ranged from 9.1 to 13.8 km during 1984 to 1988.[27] The average yearly (1984-1988) percentage of sites within 10 km of the colony was approximately 55% (range 40-65%), and the average yearly percentage of sites within 20 km of the colony was approximately 86% (range 80-95%).[23-27]

Wood storks were seen foraging in both natural and man-made wetlands (including cypress wetlands), other hardwood dominated wetlands (including blackgum, sweetgum, and red maple hardwoods), ponds, marshes, Carolina bays, drainage ditches, and other submerged depressions. The cypress and hardwood wetlands were the most frequently used foraging habitats representing 41 to 58% of the sites used during 1984 to 1988.[23-27] Prey species included fish, mostly sunfish (Centrarchidae), crustaceans (crayfish), and amphibians (tadpoles and sirens). The storks do not feed on fish less than 25 mm long as determined from samples collected from regurgitations. The densities of prey (fish > 24 mm in length, amphibians, and crayfish) at these sites during 1984 to 1988 ranged from 8.42 to 21.61 individuals per square meter, and prey biomass varied from 14.20 to 50.77 g/m^2.[27]

The Birdsville colony normally disperses in late July of each year at which time all young have fledged. The storks stay in the general vicinity of the colony until approximately late November when prey densities become too low or until the onset of freezing weather. Birdsville storks equipped with solar-powered, backpack radio transmitters were found to winter from the Okefenokee Swamp south to southern Florida. They overwintered with other storks but not with the other radio-telemetered storks from Birdsville. Two storks were located during consecutive winters. Each of these was found to overwinter in a different location during the second winter than it had during the first winter.[29] However, another stork has wintered in the same location during at least two nonconsecutive winters.

The U.S. Department of Energy's Savannah River Site

The U.S. Department of Energy (DOE) operates nuclear materials production reactors on the Savannah River Site (SRS), formerly known as Savannah River Plant, for the production of plutonium and tritium in order to meet national defense needs for nuclear weapons. Since 1953 cooling water effluents from SRS reactors have been discharged directly into several small tributary streams of the Savannah River. Starting in 1963, cooling water effluents from some of the reactors were discharged into a cooling pond. Temperatures of the cooling water effluents can be >70°C at discharge and have resulted in thermally altered aquatic habitats.[30-32]

One of the SRS streams, Steel Creek, has had an extraordinary history of perturbation.[33] Changes in the thermal loading, flows, sedimentation, and hydrologic regime in Steel Creek formed a large delta, Steel Creek Delta, at the mouth of Steel Creek where it flows into the 3020-ha cypress-tupelo Savannah River Swamp. Natural flow in Steel Creek ranged from 0.3 to 1.1 m^3s^{-1}. During 1953 to 1958, L-Reactor discharged hot water into Steel Creek and increased flows up to 1.1 m^3s^{-1}. Flows as high as 24 m^3s^{-1} were recorded from 1958 to 1963 when both L-Reactor and P-Reactor discharged into Steel Creek. P-Reactor effluents were diverted away from Steel Creek in 1963 and flows returned to 11 m^3s^{-1} levels from L-Reactor discharges. In 1968 L-Reactor was placed on standby status, and the Steel Creek ecosystem began to recover. Recovery, however, was to scrub-shrub and persistent and nonpersistent emergent plant community types instead of the original bottomland hardwood wetland habitat.[34]

An increased demand for nuclear materials led to the decision in 1980 to restart L-Reactor. According to the National Environmental Policy Act, DOE was required to prepare an Environmental Impact Statement and to offer alternatives to mitigate the impacts of thermal effluent that would be discharged into Steel Creek. DOE was able to mitigate the discharge of thermal effluent by constructing a 405-ha, once-through cooling lake, L-Lake. The preparation of the Environmental Impact Statement became more complicated, however, when wood storks were observed on the SRS.

The Steel Creek Delta Mitigation

Wood storks were followed from the Birdsville colony to the SRS, located 45 km northeast of the colony. The storks forage in the Savannah River Swamp, including Steel Creek Delta. Although the discharge below the proposed L-Lake dam would not be thermally elevated, the discharge would increase water levels too much for storks to forage in the Steel Creek Delta. Because the wood stork was proposed for listing as an endangered species in February 1983,[35] the DOE instructed the Savannah River Ecology Laboratory to study the wood stork utilization of the SRS and to deterimine any potential impacts of plant operations on the wood stork.[36]

USFWS began informal consultation with DOE in July 1983 and submitted its Biological Opinion in June 1984.[36] In its Biological Opinion, USFWS identified the Steel Creek Delta as an important foraging habitat for the wood stork. In contrast to the Florida colonies, Birdsville colony has had good reproductive success. Although 1983 was a good year for many wood stork colonies, Birdsville had the highest reproductive rate of all colonies for which data were available.

Birdsville wood storks forage on the SRS during the breeding season. Steel Creek Delta contained three of the nine sites foraged by storks in 1983. DOE estimated that 1643 pounds of food resources for storks in Steel Creek Delta would be lost because of the restart of L-Reactor. This loss was calculated to be 3.3% of the total resources needed for the Birdsville colony in 1983, equivalent to the food needed to raise nine wood storks.[36]

USFWS did not accept DOEs estimates of food loss and impacts on the colony's reproduction. The data were incomplete with only 20 days (14%) of the breeding season available to determine the importance of Steel Creek Delta on the Birdsville colony. Even an entire year's worth of data would not be enough to predict the importance of Steel Creek Delta. Some years may have fewer foraging sites available than in 1983. In 1981, for example, the Birdsville wood storks abandoned their young three to four weeks into the prefledging period. During dry years, foraging sites like Steel Creek Delta that are along more permanent wetlands would increase in importance since they would be available after other sites become dry.[36]

Of the 50 foraging sites located in 1983, only three were as far away from the colony as the Steel Creek Delta site. USFWS felt that it seemed reasonable that the Steel Creek Delta site was sufficiently important to the Birdsville colony since the wood storks repeatedly expended the energy to fly the 45 km to use the site. The failure of the colony in 1981 seemed to indicate that, even with the Steel Creek Delta available, there may be periods of insufficient food for the Birdsville wood storks.[36]

The USFWS concluded, given the available data, that the loss of Steel Creek Delta could cause the failure of the Birdsville colony during a marginal year. Finally, USFWS noted that the colony was more important than a similar sized colony in southern Florida. SInce it was the most northern and inland colony known to exist, USFWS felt it was an example of a "pioneering" group of storks that was leading the gradual movement of wood storks northward away from the failing rookeries in Florida.[36]

In order to mitigate the lost foraging habitat in Steel Creek Delta, DOE entered into an Interagency Agreement with USFWS to construct, operate, and maintain ponds for wood stork foraging habitat. USFWS stipulated the following for the mitigation to be acceptable: (a) The mitigation would be an "in-kind" replacement of the lost foraging resource and would be constructed on a site or sites agreeable to the USFWS. (b) A minimum of 12.5 ha of ponds stocked with appropriate prey would be necessary to replace the lost foraging habitat in Steel Creek Delta. (c) The ponds must be maintained as wood stork feeding areas "as long as L-Reactor operation adversely affects the foraging habitat in Steel Creek or until USFWS and DOE agree that the ponds are no longer necessary."[36] (d) Savannah River Ecology Laboratory research on the Birdsville colony will continue and annual reports will be provided on the ponds and the storks' use of them. (e) The replacement foraging ponds must be available for use when the wood storks arrive for the next breeding season (February 15, 1985) or the operation of L-Reactor must be delayed until after the breeding season is over (November 1985). Given the above conditions, USFWS's final Biological Opinion was that "ongoing and planned operations at the Savannah River Plant, including the restart of L-Reactor, are not likely to jeopardize the continued existence of the wood stork."[36]

Kathwood Lake Replacement Foraging Habitat

Design, Construction, and Operation

Kathwood Lake, located on National Audubon Society's Silver Bluff Plantation Sanctuary, Jackson, South Carolina, was selected as the site for mitigating the lost foraging habitat in Steel Creek Delta. Kathwood Lake is located about 30 km northwest of the Steel Creek Delta and about 45 km northeast of the Birdsville colony. It is similar to Steel Creek Delta in its direction and distance from the Birdsville colony. Kathwood Lake, an old mill pond, was originally built in 1850 by damming and diverting flow from a nearby creek, Hollow Creek, into a 14-ha depression. The dam broke in May 1977, and the lake drained. In September 1977, 25 wood storks were seen foraging in small pools in the lake bed, and again in 1984, three wood storks were sighted foraging in the lake basin.[37,38]

The design, construction, and operation of the Kathwood Lake replacement ponds were a cooperative effort. A Kathwood Technical Working Group was formed by representatives from the DOE, USFWS, National Audubon Society, E. I. du Pont de Nemours and Company's Savannah River Laboratory and the Savannah River Ecology Laboratory. Earlier, between 1969 and 1978, the National Audubon Society investigated the feasibility of providing fish in artificial ponds in Florida for wood storks. Eleven ponds totaling 13 ha were constructed and stocked with fish. Wood storks did forage in the ponds. However because of porous soils, it was difficult to maintain water levels in the ponds and the study was curtailed.[38]

The Kathwood Technical Working Group decided to use a similar approach to the National Audubon Society's artificial foraging ponds. Kathwood Lake was divided into four impoundments by internal levees. Ponds 1 to 4 were 4.9, 4.6, 4.8, and 1.9 ha, respectively, and totaled 16.2. Water from Hollow Creek flows successively through ponds 1, 2, 3, and 4 before returning to Hollow Creek. Water control structures in each levee make it possible to raise and lower each pond independently of the others. Construction began in August 1985 and was completed in the spring of 1986.[38]

Management

Management goals for Kathwood Lake were set by the Kathwood Technical Working Group as follows: (a) Attain a successively used foraging site for wood storks that replaces lost foraging habitat in Steel Creek Delta. (b) Attract wood storks to Kathwood Lake in numbers comparable to Steel Creek Delta prior to L-Reactor restart. (c) If possible, exceed the numbers of storks normally attracted to Steel Creek Delta by enhancing the numbers of fish available at Kathwood Lake. (d) Actively manage ponds during the first two years of operation. Thereafter, review the feasibility of working toward a more natural self-sustaining wood stork foraging system at Kathwood Lake.[39]

In order to avoid depleting the fish in Kathwood Lake during times when ample prey are available elsewhere, it was decided that fish prey in the ponds would be made available to the wood storks when they most needed extra food. During the majority of the year, the ponds would be kept at a level too deep for the wood storks to forage. Fish could reproduce and grow during these times. Three time periods were identified and prioritized when fish in Kathwood Lake would be made available to the storks. The first priority was to provide fish during the times when the Birdsville colony chicks were in greatest need of food. Both parents spend a great deal of time away from the colony foraging during this time, usually June and July. Second, fish would be provided during the postfledging period when the young storks forage for themselves but are still naive about where and how to forage. The third and last priority was to provide fish during extreme periods when other foraging sites were unavailable because of high water levels or drought.[38]

Fish would be made available to the wood storks by lowering the water level in each pond enough to make it physically possible for the storks to wade and forage, approximately 15 to 30 cm in depth. The lowered depths would also concentrate the fish and could be lowered more to maintain high prey densities as the number of fish decreased. Ponds would be made available one at at time to the storks. As prey densities became low in one pond, the next pond could be lowered to attract the storks. One of the four Kathwood ponds would not be manipulated and, in some areas would always have depths appropriate for wood stork foraging. The minimum acceptable prey densitites would be at least equal to those at the Birdsville colony, median = $15/m^2$, and higher densities, 40 to $75/m^2$, would be attempted. Only fish of the acceptable lengths, 3 to 15 cm, would be counted in the density calculations.[38]

The Kathwood ponds were stocked in their first year with sunfish and catfish that were known to be eaten by wood storks from observations at the Birdsville colony.[23] A few grass carp (*Ctenopharyngodon idella*), too large to be eaten by storks, were stocked to control aquatic plant growth. Bluegill sunfish (*Lepomis macrochirus*), brown bullhead catfish (*Ictalurus nebulosus*), and grass carp were stocked at about 20,000 fingerling bluegill, 5000 fingerling brown bullhead catfish, and 12 adult grass carp per ha. Brown bullhead catfish were selected because their eggs are less likely to be predated. The unmanaged pond received only 2500 bluegill, 250 brown bullheads, and 0 grass carp per ha. Fish reproduction and limited stocking was thought to be enough to replenish fish in the ponds from year to year. One pond was stocked with bluegill at a rate of 53,000 per ha in order to make it particularly attractive to storks and increase the chances that storks would use the ponds during the first year of operation. Wood stork decoys were placed on the ponds to help attract the storks.

Performance

Kathwood Lake has been a tremendous success. The number of wood storks that have foraged at Kathwood Lake has far exceeded all expectations since they first arrived in July 1986. The maximum group size of wood storks observed together on Kathwood Lake during an entire day has increased each year from 97 wood storks to 151, 212, and 223 wood storks during 1986 to 1989, respectively (Figure 6).[25-27]

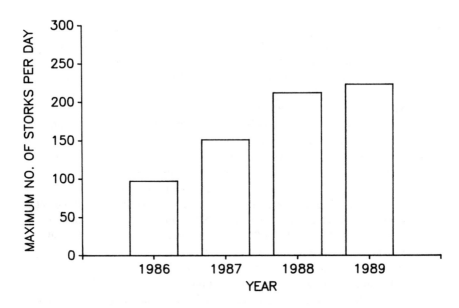

Figure 6. Maximum number of wood storks observed in group during an entire day at Kathwood Lake from 1986 to 1989.

The total stork usage at Kathwood Lake per year has also increased each year from 3074 stork-days use (maximum group size per day summed for all days) in 1986 to 3338 stork-days use in 1987, 4193 stork-days use in 1988, and 4612 stork-days use in 1989.[25-27]

The density estimates of fish in Kathwood Lake varied from 0.6 to 102 fish per m^2 during 1986 to 1988. The fish densities were kept adequately high to feed wood storks weeks past their postfledging critical stage by continually decreasing pond water levels. Lengths of fish were equally acceptable and varied from 21.9 to 82.9 mm during 1986 to 1988. Natural fish reproduction in Kathwood Lake was adequate enough that minimal restocking was done only as a precaution. Up to ten other species of fish besides those stocked plus grass shrimp (*Palaemonetes paludosus*) provided additional prey for the storks by inhabiting Kathwood Lake from populations in Hollow Creek.[25-27]

Much of the success of the Kathwood Lake mitigation has been due to a well-defined goal, excellent cooperation among the organizations involved, and continued research in support of the mitigation effort. The DOE has been exemplary in its fulfillment of its obligations under the ESA. The total cost of the 5-year mitigation effort was about $3.3 million. Private landowners, public land managers, and several National Wildlife Refuges have requested further information from the USFWS on the Kathwood Lake design and management for wood storks elsewhere.[40] During the Kathwood Lake mitigation, the USFWS completed its Wood Stork Recovery Plan.[41] The rigorous application of this plan, along with continued research, and efforts similar to the Kathwood Lake effort to protect the wood stork throughout its range, should play a vital role in the successful recovery of this endangered species.

UNRESOLVED QUESTIONS

The DOEs restart of the L-Reactor was very controversial. There were debates over the need to restart the reactor, the cost of restart, and impacts to the environment. The loss of foraging habitat for an endangered species, the wood stork, fueled the debates. Proponents for the wood stork requested that the ESA be carefully followed. Others criticized the wood stork mitigation as too costly, especially when it was not certain that Kathwood Lake would be found and used by the storks. Nevertheless, the cost of the wood stork mitigation was relatively low when compared to the mitigation cost for the rest of the L-Reactor restart.

Kushlan[42] reviewed the USFWSs Recovery Plan for the wood stork and identified several concerns. The plan gave the current United States population of the wood stork as 5000 pairs. This was close to the 6000 pairs that the plan identified as acceptable for down-listing from endangered to threatened. Furthermore, Kushlan noted arguments that the wood stork should not have been listed as endangered, and perhaps not even threatened since there were large populations of the species outside the United States. If may have been more realistic to identify the southern Florida population of wood storks as endangered. This would have concentrated research and management efforts to the area where the major wood stork problems reside. Finally, Kushlan felt that enough research had already been done on the southern Florida populations, that more management practices could begin immediately than those proposed in the Recovery Plan. He reminded conservationists that "the recovery process is a thoroughly political one."

Many questions remain regarding the magnitude and type of effort that should be expended on protecting endangered species. For example, should the focus be on the preservation of individual endangered species or on the preservation of habitat? It is often very expensive to bring about the recovery of a species after it is already in danger of extinction. Perhaps money and effort could be better spent on preserving habitats and, thus, indirectly helping a larger number of species not yet endangered. It may be, however, much easier to rally support for the protection of endangered species than for habitats. Single, sometimes popular, species can become the focus of concern and funding. Furthermore, protection of specific endangered species often results in the protection of habitats and other species.

SUMMARY AND CONCLUSIONS

As in the case of the wood stork, there are success stories where seemingly unresolvable conflicts can be adequately resolved without furthering the decline of a species. Nevertheless, the current rate of species extinction is unacceptably high.

The exploding human population and its unrestricted exploitation and destruction of habitats are the major causes of the extinctions. Protection of species threatened with extinction is not a choice but a necessity if the collapse of the earth's ecosystem and mass extinction of species, including humans, are to be avoided.

ACKNOWLEDGMENTS

The authors are very grateful to Kathryn J. Bailey for her assistance with graphics, preparation of this chapter, and review comments. The authors also thank Patricia J. West for her critical review and helpful comments. Numerous technicians have contributed over the years to the Savannah River Ecology Laboratory stork project including A. L. Bryan, Jr., S. L. Coe, F. C. Depkin, T. L. Gentry, L. C. Huff, S. D. Jewell, W. B. Lee, D. E. Manry, L. S. McAllister, K. L. Montgomery, L. A. Moreno, M. A. Rubega, D. J. Stangohr, W. J. Sydeman, N. K. Tsipoura, J. M. Walsh, B. E. Young, and D. P. Young. This chapter was supported by Contract DE-AC09-76SROO819 between the U.S. Department of Energy and the University of Georgia's Savannah River Ecology Laboratory.

REFERENCES

1. Ehrlich, P. R. "The Loss of Diversity, Causes and Consequences." in *Biodiversity*, E. O. Wilson, Ed. (Washington, D.C.: National Academy Press, 1988), pp. 21-27.
2. Thompson, W. S., and D. T. Lewis. *Population Problems* (New York: McGraw Hill Book Company, 1965), p. 384.
3. "1986 Demographic Yearbook, Thirty-eighth Issue." United Nations (1988), p. 147.
4. "Statistical Abstract of the United States, National Data Book and Guide to Sources, 103rd Ed." U.S. Bureau of the Census (1982), p. 856.
5. Zachariah, K. C., and M. T. Vu. *World Population Projection, 1987-1988 Edition, Short- and Long-Term Estimates* (Baltimore, MD: Johns Hopkins University Press, 1988), p. xviii.
6. Nilsson, G. *The Endangered Species Handbook* (Washington, D.C.: The Animal Welfare Institute, 1983), pp. 210-213.
7. Wilson, E. O. "The Current State of Biological Diversity." in *Biodiversity*, E. O.Wilson, Ed. (Washington, D.C.: National Academy Press, 1988), pp. 3-18.
8. Myers, N. "Tropical Forests and Their Species, Going, Going ...?" in *Biodiversity*, Wilson, E. O., Ed. (Washington, D.C.: National Academy Press, 1988), pp. 28-35.
9. Original author of analogy unknown.
10. Drabelle, D. "The Endangered Species Program," in *Audubon Wildlife Report 1985*, A. S. Eno, Proj. Dir., R. L. Di Silvestro, Ed. (New York: The National Audubon Society, 1985), pp. 73-90.
11. *Endangered Species Technical Bulletin*, U.S. Fish and Wildlife Service, Department of the Interior, Washington, D.C., (1976-1989).
12. *Endangered Species Technical Bulletin*, Vol. 14, No. 4, (Washington, D.C.: U.S. Fish and Wildlife Service, 1989).

13. Reffalt, W. C. "United States Listing for Endangered Species, Chronicles of Extinction?" *Endangered Species Update* 5(10):10-13 (1988).
14. Kahl, M. P. "An Overview of the Storks of the World" *Colonial Waterbirds* 10(2):131-134 (1987).
15. Ogden, J. C., D. A. McCrimmon, Jr., G. T. Brancroft, and B. W. Patty. "Breeding Populations of the Wood Stork in the Southeastern United States" *The Condor* 89: 752-759 (1987).
16. Endangered and Threatened Wildlife and Plants; U.S. Breeding Population of the Wood Stork Determined to be Endangered," 50 CFR Part 17, *Federal Register* 49(40):7332-7335 (1984).
17. Luthin, C. S. "Status of and Conservation Priorities for the World's Stork Species" *Colonial Waterbirds* 10(2):181-202 (1987).
18. Ogden, J. C., and S. A. Nesbitt. "Recent Wood Stork Population Trends in the United States" *Wilson Bull.* 91(4):512-523 (1979).
19. Kushlan, J. A., and P. C. Frohring. "The History of the Southern Florida Wood Stork Population" *Wilson Bull.* 98(3):368-386 (1986).
20. Ogden, J. C., and B. W. Patty. "The Recent Status of the Wood Stork in Florida and Georgia" Technical Bulletin WL5, in *Proceedings of the Nongame and Endangered Wildlife Symposium* (Athens, GA: Georgia Dept. of Natural Resources, Game and Fish Division, 1981), pp. 97-101.
21. Sharitz, R. R., and J. W. Gibbons. "The Ecology of Southeastern Shrub Bogs (Pocosins) and Carolina Bays: A Community Profile," FWS/OBS-82/04, U.S. Fish and Wildlife Service, Division of Biological Services, Department of the Interior, Washington, D.C. (1982), 93 pp.
22. Meyers, J. M. "Wood Storks of the Birdsville Colony and Swamps of the Savannah River Plant," Savannah River Ecology Laboratory Report SREL-15 (1984), 48 pp.
23. Coulter, M. C. "Wood Storks of the Birdsville Colony and Swamps of the Savannah River Plant, 1984 Annual Report," Savannah River Ecology Laboratory Report SREL-20 (1986), 110 pp.
24. Coulter, M. C. "Wood Storks of the Birdsville Colony and Swamps of the Savannah River Plant, 1985 Annual Report," Savannah River Ecology Laboratory Report SREL-23 (1986), 126 pp.
25. Coulter, M. C. "Wood Storks of the Birdsville Colony and Swamps of the Savannah River Plant, 1986 Annual Report," Savannah River Ecology Laboratory Report SREL-31 (1987), 242 pp.
26. Coulter, M. C. "Wood Storks of the Birdsville Colony and Swamps of the Savannah River Plant, 1987 Annual Report," Savannah River Ecology Laboratory Report SREL-33 (1988), 266 pp.
27. Coulter, M. C. "Wood Storks of the Birdsville Colony and Swamps of the Savannah River Plant, 1988 Annual Report," Savannah River Ecology Laboratory Report, SREL-37 (1989).
28. Tate, A. L., and R. L. Humphries. "Wood Storks Nesting in Jenkins County, Georgia" *Oriole* 45(2&3):34-35 (1980).
29. Comer, J. A., M. C. Coulter, and A. L. Bryan, Jr., "Overwintering Locations of Wood Storks Captured in East-Central Georgia" *Colonial Waterbirds* 10(2):162-166 (1987).
30. Sharitz, R. R., J. E. Irwin, and E. J. Christy. "Vegetation of Swamps Receiving Reactor Effluents" *Oikos* 25:7-13 (1974).
31. Gibbons, J. W., and R. R. Sharitz. "Thermal Ecology: Environmental Teachings of a Nuclear Reactor Site" *Bioscience* 31:293-298 (1981).

32. McCort, W. D. "Effects of Thermal Effluents from Nuclear Reactors" in *Environmental Consequences of Energy Production: Problems and Prospects,* S. K. Majumdar, F. J. Brenner, and E. W. Miller, Eds. (Easton, PA: The Pennsylvania Academy of Science, 1987), pp. 387-401.
33. Smith, M. H., R. R. Sharitz, and J. B. Gladden. "An Evaluation of the Steel Creek Ecosystem in Relation to the Proposed Restart of L-Reactor," Savannah River Ecology Laboratory Report SREL-9, NTIS, Springfield, Virginia (1981).
34. Dunn, C. P. and R. R. Sharitz. "Revegetation of a *Taxodium-Nyssa* Forested Wetland Following Complete Vegetation Destruction" *Vegetation* 72:151-157 (1987).
35. "Endangered and Threatened Wildlife and Plants; Proposed Endangered Status for the U. S. Breeding Population of the Wood Stork," CFR Part 17, *Federal Register* 48(40):8402-8404 (1983).
36. Parker, W. T. "Wood Stork Biological Opinion," (Asheville, NC: U.S. Fish and Wildlife Service, 1984).
37. Connelly, D. M. National Audubon Society, Silver Bluff Plantation Sanctuary, Jackson, SC. Personal communication (1987).
38. Coulter, M. C., W. D. McCort, and A. L. Bryan, Jr., "Creation of Artificial Foraging Habitat for Wood Storks" *Colonial Waterbirds* 10(2):203-210 (1987).
39. McCort, W. D., M. C. Coulter, A. B. Gould, Jr., H. E. Mackey, Jr., N. A. Murdock, J. C. Ogden, and D. M. Connelly. "Kathwood Lake Wood Stork Management Plan," Savannah River Ecology Laboratory Report (1986).
40. Murdock, N. A. "Creation of Artificial Foraging Habitat for Wood Storks" *Endangered Species Technical Bulletin* 12(1):4,16 (1987).
41. Bentzien, M. M., "Recovery Plan for the U.S. Breeding Population of the Wood Stork," U.S. Fish and Wildlife Service, Jacksonville, FL (1986).
42. Kushlan, J. A. "Recovery Plan for the U.S. Breeding Population of the Wood Stork" *Colonial Waterbirds* 10(2):259-262 (1987).

CHAPTER 12

The Savannah River — Past, Present, and Future

Ruth Patrick

From a geological standpoint, the Savannah River (Figure 1) is one of the older rivers in the United States. Its economic importance has grown since it was first discovered by DeSoto in 1540. However, accounts say he went up the Savannah River but did not establish any settlements. The first settlement of record was Savannah Town, established in 1680 as an Indian trading station. Many encounters, both friendly and unfriendly, occurred between the Indians, mainly the Cherokee tribe, who were trading with the white people. Unfriendly relations resulted in the Yemassee War (1715 to 1718). During this period, Colonel Morris Moore and his 300 men entered the area around Augusta and convinced the Cherokees to live in quasi-peaceful conditions with them. At that time, Savannah Town, which had its site in North Augusta, was renamed "Fort Moore" in 1717. In 1736, General James Oglethorpe changed the name of Fort Moore to Augusta.

After 1717, small towns sprung up along the shores of the Savannah River, such as Andersonville at the mouth of the Seneca River, and Petersburg, Lisbon, and Vienna near the mouth of the Broad River. Plantations were the early form of agriculture in the Savannah River Basin, and the main crops were cotton and tobacco. The first settlements in the coastal plain were by Swiss-Germans in 1730. Led by Jean Pierre Purry, they settled on the great Yemassee Bluff about 22 miles up river from the ocean. These hard working immigrants tried economics based on indigo, silk, hemp, cotton, and wine, but the unhealthy climate proved their undoing. Many died and others departed the colony.

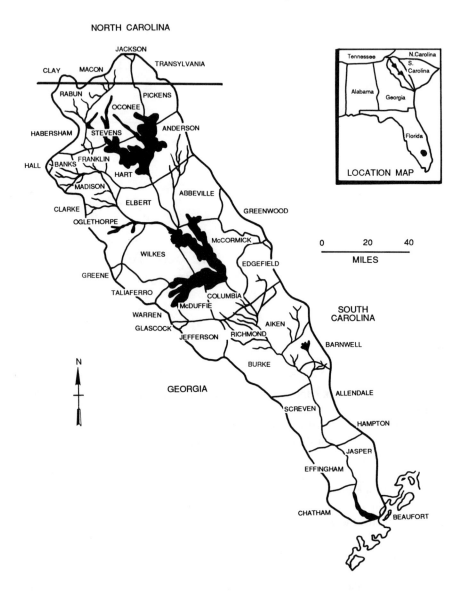

Figure 1. Savannah River drainage basin.

In the 1740s and 1750s, settlers gradually pushed into Barnwell, Aiken, and Edgefield counties. In 1759, Patrick Calhoun and a few Scotch-Irish families settled on Long Cane Creek in Abbeyville County.

The first railroad, which was at that time the longest railroad in the United States, was built between Hamburg and Charleston in 1833 to carry cotton from the plantations in the Piedmont down to the port of Charleston on the coast. In

1846, Graniteville Cotton Factory was built. After the Civil War in 1865, cotton growing transformed the antebellum South, and Eli Whitney invented the cotton gin on the plantation of Mrs. Nathaniel Greene. Textile mills developed toward the end of the 19th century and flourished in the beginning of the 20th century. After the Civil War, slave labor was replaced by crop sharers, and in many cases, the productivity of these plantations decreased, partially because of the poor economic conditions of the South.

The increase in the boll weevil at the end of World War I and the transportation provided by automobiles brought about the decline in the cotton economic prosperity of the watershed. Between World Wars I and II, South Carolina and Georgia were not very productive, and the population in the coastal plain declined. It did steadily increase in the Piedmont and in the mountain regions.

World War II brought a new industrial age to the Savannah River Basin with the introduction of hydroelectric power. This enabled larger industries to profitably manufacture their products in the Savannah River watershed. Many dams have been built along the Savannah River which have had great impact on the Savannah River. From the riverine standpoint, they have brought about deterioration of habitats for aquatic life. From a socioeconomic standpoint, they have brought about prosperity in the watershed. Accompanying the increased industrialization has been the dredging of the Savannah River by the Corps of Engineers. A 9-ft channel is maintained throughout most of the Savannah River and is deepened to 32 ft in the harbor. Total hydroelectric power potential in the Savannah River Basin is estimated to be about 2000 MW. Today, about one half that potential for hydroelectric power has been realized. The use of the Savannah River Basin is changing rapidly. The greatest development so far has been in the Piedmont and around the estuary. The coastal plain is not very well developed and will probably, in the next 50 years, be the rapidly growing area.

The Savannah River, which is the heart of the Savannah River Basin, rises in the Blue-Hill Mountains, flows through the Piedmont and the coastal plain, and enters the Atlantic Ocean at Savannah. As seen in Figure 2, the profile of the river, when plotted with an elevation drop against the distance from the ocean, forms more or less a hyperbola. The steep gradient areas are the ones in the mountains and foothills, and the more gentle sloping area is the region of the Piedmont where the river flows over consolidated soils. At Augusta, or slightly above Augusta, is the fall line where the Piedmont merges with the coastal plain. The rocks disappear from the riverbed, and instead of the river having a consolidated bed that is wide and shallow, the bed shifts to an unconsolidated form with a relatively narrow channel with steep banks. The area of very low or no gradient in the estuary is where fresh- and salt water mix.

The energy of the flowing water is dissipated against the bed and the sides of the channel. Since the bed is unconsolidated, the river is very deep in many places. Strong meanders develop in the river course in order to dissipate the energy of the flowing water. These meanders are very important to the aquatic life of the river. The sediment load of the Savannah River is large, and the photosynthetic zone is relatively shallow. In the main channel, the only areas that are shallow enough for

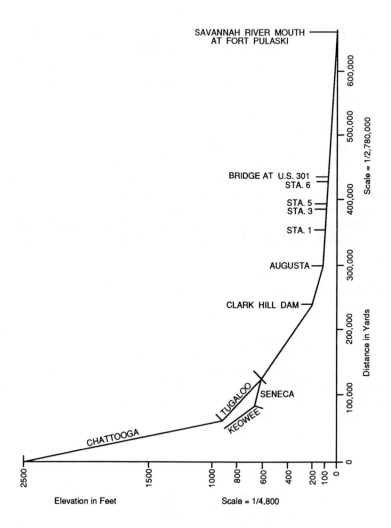

Figure 2. Savannah River profile — surface gradient.

benthic photosynthetic species and the accompanying animal life are on the point bars. Sloughs and oxbows are common along its course. These oxbows are very important for the biology of the river for it is here that the sediments settle out of solution, the algae and invertebrates feed and reproduce, and the fish come to feed. Without these oxbows, the biological productivity of the Savannah River would be greatly reduced. At present, there is a trend to straighten the channel and to cut out the oxbows in order to make a more efficient river for transportation. This is counter to the suitability of the Savannah River as a habitat for fish and associated species. If the dredging and isolation of oxbows continue, a reduction in species of fish, insects, and other aquatic life will occur.

The estuary, as stated previously, is the region of the freshwater-salt water interface. At this interface, the colloidal fraction of the sediments, which often carries metals, radioactive materials, and organic compounds, settles out of solution, thus producing an environment that is not very suitable for many forms of aquatic life. The salinity wedge of the ocean extends up into the estuary, and the extent of this salt water tongue is dependent upon the flow of the river.

The Savannah River estuary is divided into a front and a back channel a short distance above its mouth. A great deal of industrial development, particularly sugar refining, paper, and various types of manufacturing activities, has occurred in the front channel of the estuary. The estuary, however, is also an important breeding and/or nursery ground for striped bass, shrimp, or crabs. In the early days, shad and sturgeon were frequent migrants in the Savannah River. Over time, specialized strains of shad have developed, which spawn in the tributaries of the Savannah River and remain there until they are mature, i.e., the young of the year do not migrate to the sea as is typical of most shad populations. The mouth of the estuary is surrounded by marshy grasslands that are very important as breeding grounds and also as sinks for various pollutants that may have entered the river water.

The Savannah River, since the time of the practice of agriculture by early settlers, has had a very heavy sediment load that is mainly picked up from the erosion of land in the Piedmont. The Broad River carries the largest sediment load. The extent of this erosion greatly increased for over 50 years. In 1850, the average county in the lower Piedmont was one third woodland, one third cultivated, and one third so badly eroded that it had little use.

Land management in the 20th century has reduced this sediment load, and very little sediment is picked up in the coastal plain. The Savannah River is a soft water river, being very low in alkaline metals. As it enters the coastal plain and the region where swamps are most prolific, the humates of the swamp enter the river. These humates have an oxygen demand and also increase the acidity of the river. Thus, the fauna and flora are a mixture of typically soft water species and those found in waters high in humates. In the estuary, the fauna and flora are a mixture of fresh- and salt water species. Thus, today, the river supports a fairly natural diversity of aquatic life throughout its course. There are some places where pollution is heavy and the aquatic ecosystems show signs of degradation due to these types of perturbations.

However, in general, the Savannah River is in better condition than it was when studies began in the region around the Savannah River Plant (Figures 3 to 7). At that time, sewage treatment was minimal, and many toxic chemicals entered the Savannah River through Horse Creek. The prevention of large fluctuations in water levels has favored some species, but has eliminated the nursery areas of flood plain ponds and subsequent seeding of the river of species that lived in these ponds.

Because of state laws and the Federal Water Pollution Control Acts, early pollutants have been reduced. However, with the expanding population predicted in the future, this pollution load may again increase in the form of excess nitrates, phosphates, and trace metals caused by the types of nonpoint sources entering the

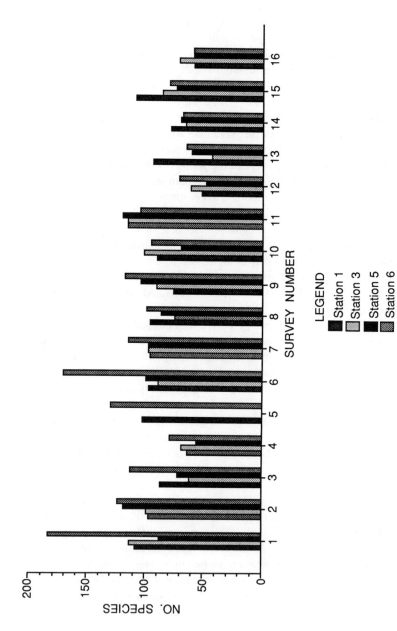

Figure 3. Algae, 1951 to 1976, above (Station 1) and below the Department of Energy site on the Savannah River.

THE SAVANNAH RIVER—PAST, PRESENT, AND FUTURE 143

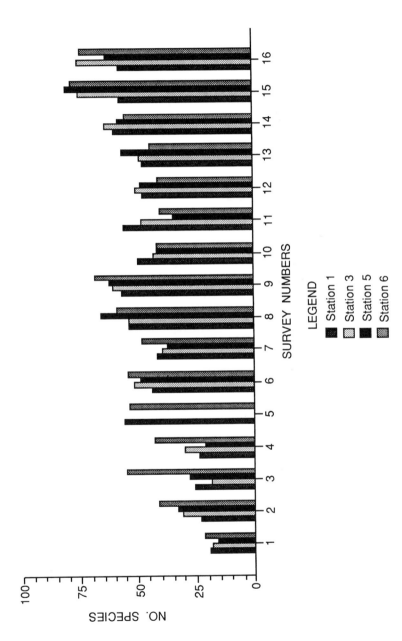

Figure 4. Protozoa, 1951 to 1976, above (Station 1) and below the Department of Energy site on the Savannah River.

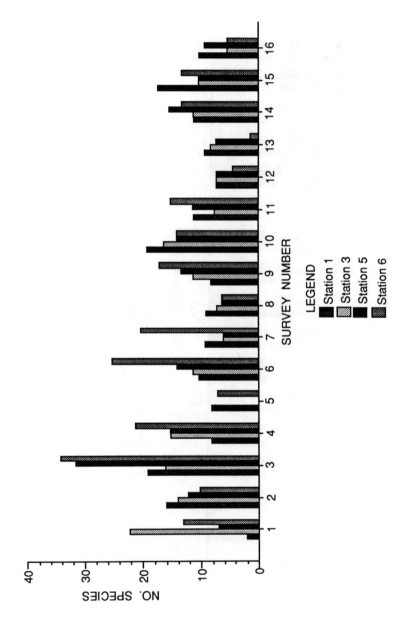

Figure 5. Invertebrates, 1951 to 1976, exclusive of insects above (Station 1) and below the Department of Energy site on the Savannah River.

THE SAVANNAH RIVER—PAST, PRESENT, AND FUTURE   145

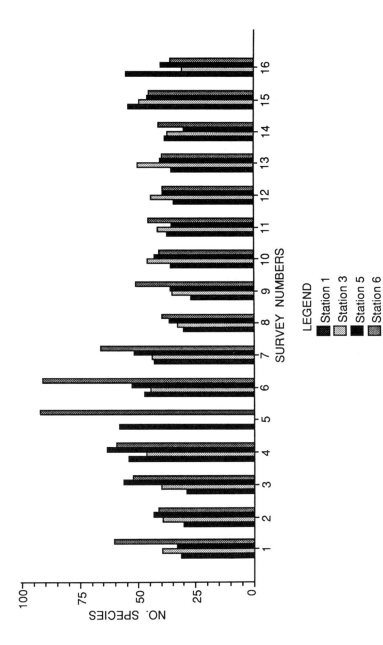

Figure 6. Insects, 1951 to 1976, above (Station 1) and below the Department of Energy site on the Savannah River.

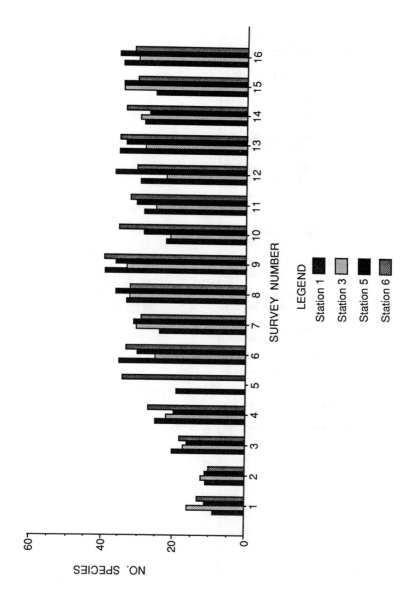

Figure 7. Fish, 1951 to 1976, above (Station 1) and below the Department of Energy site on the Savannah River.

river, as well as the type of sewage treatment. Eutrophication due to excess nitrates and phosphates will probably be more evident in the river than it is today.

The population will probably grow at a rapid rate in this area. The best available data are for the central Savannah River area, which is a 13-county area in South Carolina and Georgia. The fastest growing category during the next 2 decades will be the 30- to 44-year-old age group, followed by the 45 to 59, and the 60+. The slowest growing age group will be from 0 to 14 and 15 to 29. The 30 to 44 age group is predicted to increase 70.8% between the years 1980 and 2000. The 45 to 59 age category will increase 66.2%. Much of this increase will occur between the years 1990 and 2000. The 60+ age category is also projected to show strong signs of growth, increasing about 43.4%. The age categories under 30 years are also projected to grow during the next 2 decades, although at a slower rate than those 30 or above. The 0 to 14 age category will increase about 28.5%. The group showing the slowest growth rate during the next 2 decades will be the 15 to 29 year age category with the growth rate of 8.9%. Although the slowest growing age category, this group contains the largest segment of the population at the present time.

The older age groups will mainly live in the mountainous and Piedmont areas, whereas the younger age groups will mainly live around the industrial development in and around Augusta and in the coastal plains area. At present, the greatest amount of industrial activity is manufacturing, which is 25% of the industrial activity; service industries, which are 31%; and retail trade, which is 15% (Figure 8). Construction is 7% and transportation is 6%. Undoubtedly, with the growth of the older population, more recreation will be demanded, and this will increase the industries (both service and production) that are associated with recreational activity.

These predictions of increased population may be low considering the national figure that by the early part of the next century, 70% of the population of the United States will live within 50 to 100 mi of the coast. Since South Carolina and Georgia are relatively underpopulated and have desirable winter climates, the growth of these areas may be much larger than is anticipated.

This heavier usage of the watershed will undoubtedly result in an increase in nonpoint sources of pollution unless they are controlled. At present, only a few state and federal laws address, even tangentially, the problem of nonpoint sources.

There are many questions facing the inhabitants of the Savannah River Basin as to what they wish the Savannah River and its associated basin to produce in the long run. What kind of socioeconomic activity is wanted? What kind of a river is desirable? The increase of dams, which are now very significant in the Savannah River, has greatly altered the recharge areas of the shallow water aquifers, particularly in the Coastal Plain. This means that water use in this area will not develop to its full potential because of the creation of the dams. The dams supply hydroelectric power and increase the recharge of some of the groundwater aquifers in the Piedmont, but this recharge may or may not influence the recharge of the shallow water aquifers in the Coastal Plain. Since these aquifers are the greatest potential source of potable water, this may become a serious problem for population growth in the Coastal Plain. The recharge of these aquifers must be determined, as well as

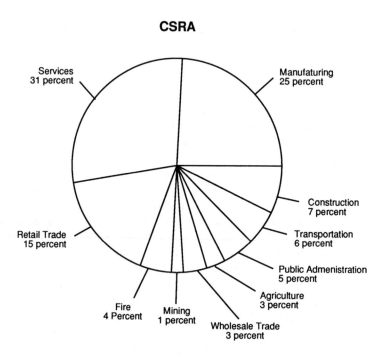

Figure 8. Industrial activity in the environs of Augusta.

those of the deeper confined aquifers. Then, a plan must be developed for water use that is in line with the recharge potential of the aquifers. Since the Savannah River is utilized for potable water as well as manufacturing, the quality of water must be maintained for drinking water purposes. Although sewage treatment may mitigate many of the problems of sewage, it may also increase the nitrates and phosphates in the river water and associated pollution problems (e.g., eutrophication).

In future planning, sludge from these municipal treatment plants must be carefully considered. Spreading sludge on the land has been an option. In small proportions, this may be a suitable solution, particularly in the Piedmont. However, in the Coastal Plain where the soils are sandy, the nitrates and other soluble compounds from the sewage will penetrate rapidly through the sandy soils into the unconfined aquifers. These sludges should not be placed in wetlands, which are recharge areas for unconfined aquifers.

Natural resources are limited. In the future, areas may be declared as full, i.e., that the carrying capacity of the area will only support a given number of people, and when that number is reached, no more people can live in the area. Likewise, the type of industry may have to be regulated. At present, there is a great deal of consumptive use of water, particularly by utilities in the Coastal Plain. That, together with the dams in the upper river systems (which prevent flooding and recharge of groundwater in the lower reach of the river), may bring about a limitation for this type of water use. It is very important, as plans for the future are implemented, that the use of the estuary by industries and utilities and the riverine system for navigation and for pollution disposal be balanced against the needs of recreation and the maintenance of viable fisheries and quality of water suitable for contact recreation.

Plans for the Savannah River and its watershed should be developed in line with future population demands. What will the future population wish? How should development take place? Air, water, and land form one interacting environment, and therefore any plan for use of water or disposal of wastes must be examined as to its ultimate fate as well as its immediate effects.

CHAPTER 13

Impacts of Management Decisions on Environmental Issues of the Savannah River

Paul B. Zielinski

INTRODUCTION

The Savannah River drains land from the mountains to the Atlantic Ocean, a distance of over 300 mi. The river contains a number of impoundments between the mountains and the fall line: three private dams with two having pumped storage and three public dams with one having pumped storage capability. The private dams are owned by Duke Power and consist of Keowee Dam (the lowest), and Jocassee and Bad Creek dams, containing pumped storage. The public dams are the Strom Thurmond, Richard B. Russell, and Hartwell dams, with the Russell dam having pumped storage capability. The Strom Thurmond Dam was originally constructed as the Clark Hill Dam.

Since the lower Savannah River is highly susceptible to flooding and since there was a need to maintain a navigation channel between Augusta and Savannah, Clark Hill Dam was built. Power generation and silt control were also functions of the new lake, and recreation and storage of water became additional benefits of this impoundment. Subsequently, Hartwell Dam and the Richard B. Russell projects were built, justified by the same objectives. The Richard B. Russell Dam was controversial because it placed an impoundment over a sloped river used for boating and the impact of the pumped storage capability on water quality was unknown. Water stored in these dams was also used as a cooling fluid for nuclear materials production reactors at the Savannah River Site.

Conditions have changed since the time of construction of the dams. Navigation is no longer considered essential.[1] The droughts of the 1980s necessitated the Corps of Engineers (COE) consider and adopt a drought plan for COE-managed lakes, which is in effect today. The drought plan called for a reduction in the quantity of flow releases into the lower Savannah River to save water. Recreation users were disappointed in the low lake levels, and recreational activities decreased. Concerns were voiced over the quantities of stored water available for water supply with the continuation of drought conditions. While the economics of power generation revenues dictated the regulation of lake levels previously, the need for water conservation curtailed the generation of electricity, much to the concern of the power companies, since hydropower was less expensive and more readily available than other forms of generated power during peak load periods. The dedication of the Chattooga River as a federal scenic river precluded any further attempts to complete earlier planned projects for that branch of the river system. Perhaps the greatest difference between considerations today and previous ones (as in 1975[2]) is the emphasis on the environment as a major thrust. Population and industrial growth have placed new stresses on guaranteed quantities of stored clean water.

Water quality in the Savannah River Basin is generally good, with some particularly unique problems. Studies have shown an increase in concentrations of mercury compounds in all of the new impoundments, but these concentrations declined after a period of time.[3] PCB contamination in large concentrations is a problem in the upper reaches of the Twelve Mile Creek section of Lake Hartwell, and the migration and extent of these concentrations have been studied.[4] Releases from the dams during the time of year when the lakes are stratified have been deficient in dissolved oxygen.[2] Manganese is a problem in Lake Russell.[5] Turbidity due to soil erosion is a problem in the lakes immediately following heavy rainfall periods.[2]

The entire Savannah River system can be managed by the COE, Duke Power, and related agencies. The management options and the impact of actions upon the environment must be considered.

ISSUES AND IMPACTS

Perhaps one of the prime reasons for placing an impoundment in steep terrain above the fall line is to prevent flood waters from reaching the lower Savannah River. In the full bank flow in this segment without flooding, a maximum capacity of flow rate is about 40,000 cfs, according to an informed source. The COE determined that a flow rate minimum of 5800 cfs is needed to maintain a 9-ft depth between Augusta and the New Savannah Bluff Lock and Dam.[1] Early flood records for the years 1916, 1929, 1951, and 1976 (taken at Augusta) give depths of flow of about 40 ft above average depth, which yield peak flows in the range of well in excess of 100,000 cfs.[6] The lower Savannah River is highly susceptible to flooding. More exact numbers may be available in the near future when the South Carolina Water Resources Commission staff has its Savannah River Basin math model operational.[7] The Commission is developing a comprehensive computer model to simulate the water budget in the entire Basin. Previous records show that the storms generating flood flows on the order of record flood capacity tend to occur during the hurricane season. This requires that adequate flood storage volume be provided within the impoundments during the storm and rainy seasons.

For most situations, high lake levels are not a problem except when peak flood storage volume is needed within the reservoirs. One of the problems in containing peak flows within the reservoirs is retaining all of the silt within the reservoirs.

Another impact on the lower Savannah system is the change in salinity in the wetlands region from salt water to freshwater. After the COE experienced severe drought several times during the 1980s when the reservoir levels were below the desired rule curve elevations, the COE adopted (March 1989) an operational scheme called the "Savannah River Basin Drought Contingency Plan"[1] to deal with drought periods. Some items could improve the effectiveness of the plan. First, the condition that defines a drought should be identified. Defining the exact beginning of a drought is difficult, and if the timing to identification could be shortened, it would conserve more stored water by curtailing water uses and releases. The COE uses lake levels as an indicator, but lake levels alone are not adequate to define a drought, particularly one of long duration. However, lake level elevations taken in conjunction with other information, such as rainfall deficit during the water year and soil moisture content as defined by the National Weather Service (called the Palmer Index), would be closer to the actual condition. In 1987 there was a full reservoir at the start of summer, yet farmers were suffering severely from a drought before the COE defined a drought condition and changed its scheme of releases and power generation. Research is needed to develop an accurate set of criteria, but the result will be a trigger mechanism that would be more accurate and water conservative. Rainfall and soil moisture records are easily obtained from the National Weather Service and the state climatologist. One problem the COE has in defining a drought condition is that it must first give public notice before changing its downstream releases. This is another necessary delay in the process of responding to a drought and further explains why a sensitive set of criteria are necessary!

Another issue related to drought is the level of water in the reservoir at all times during the time when a drought situation is declared. Rather than follow a set of rule curves, which is a plot of reservoir elevations for each month of the year, the COE could adopt another scheme, such as developing and using a computer math model simulation of the entire Savannah River Basin. This model would include rainfall-runoff models and flood routing through the basin reservoirs. This idea is not new; it is not prohibitively large; and it is actually being developed by others at this time. As the agency in charge of reservoir operations, the COE should adopt this latest technology so that optimum use is made of the stored waters. When the model is operational, runs could be made on a frequent basis. During a drought period, the user public becomes very concerned about the wise use of its resource. There is concern about quantities of water stored and curtailed ability to make recreational use of the lake. Rather than following rule curves, other rules could be followed that optimize the use of the water and maintain the highest water levels available throughout the year. These could be the guiding principles: generate no power unless water is available, and release no water past the dam unless it is necessary to do so for safety or to generate power. In normal times, the releases should satisfy all downstream users (such as the Savannah River Site), maintain water quality in the lower Savannah basin including the wetlands, and, when necessary, continue navigation on the river below Augusta. The nuclear materials production reactors require large quantities of cooling water.

Hydropower generation lowers lake levels. The optimum time to produce power is generally peaking power during heating and cooling seasons. When power is generated and water is released, passed water cannot be recovered to use for other purposes without pumped storage. At the present time, the Russell facility has been equipped for pump storage, but the facility remains unused. Large-scale power generation is in conflict with full lake pools that are required for recreation, water supply, water quality, and cooling water. What are the environmental impacts caused by low water levels? Exposed mud banks are unsightly and may deprive lakeside residents of the use of the facility. Docks are left high on the banks and, unless they are properly built, may suffer structural damage when water levels fall. The lowered water levels and mud banks deplete the property values around the lakes. Those using the lakes during depressed lake levels may encounter hazards such as stumps that can damage boats.

Releases from the dams during power generation does create some problems. The water is low in dissolved oxygen when the lakes are stratified and, in the case of releases from Lake Russell, also contains a toxic form of manganese.[5] An attempt to improve the power generation tailwater quality at Russell Dam with the installation of oxygen injection equipment has been made by the COE.

A severe demand on groundwater aquifers, principally the tertiary or uppermost aquifer, in the area of Savannah is causing a serious problem with salt water intrusion in the Beaufort, Hilton Head, and the Savannah region.[8] Several solutions have possibilities in resolving the problem of declining groundwater levels. The most direct solution involves a discontinuation of pumping groundwater and switching to a surface water source (for most large water users that would be pumping from the Savannah River). The groundwater quality in this area is good; thus, those switching to Savannah River water must build and maintain a water treatment plant system at a significant expense. Since the salt water intrusion problem is somewhat removed from the location of the highest groundwater use, getting users to change to a Savannah River surface water source will be difficult. Another aspect of this situation involves the Savannah area industries that use groundwater. Since the groundwater temperature is significantly cooler than surface water, it is more effective and efficient as a cooling source. Pumping energy requirements would be significantly greater for river water cooling than for groundwater.

The wetlands in the Savannah Wildlife Refuge are sensitive to changes from freshwater to salt water. During times of normal releases downstream from Lake Thurmond, salt water encroachment is not a problem. In times of low flow releases, the salt water moves farther upstream in the estuary and changes the habitat from fresh to salt water. Sudden changes in salinity create problems for the plants and biological life. Salinity levels are monitored in the Savannah estuary; however, changes are likely during drought periods. This valuable resource must be protected.

Another biological activity that requires reasonably high water in the springtime is the spawning of fish. During drought periods when lake levels are low, some of the best spawning habitats are above the waterline and unavailable. Past drought periods with increased lake levels should increase spawning and fish populations due to an increased habitat of grasses that have grown on banks and stream bottoms during dry periods.

Water is currently diverted from the Savannah River Basin. When diversions occur, they deprive the originating basin of the use of that water for any purpose, forever. Diversion of water can only be noncontroversial when the diversion is removed from excess flow into a basin. In times of drought, how are the rights of a water borrower handled? How much water can be diverted from a basin? How does the increased flow affect the receiving stream? These questions must be considered in any management scheme for the Savannah River Basin.

CONCLUSION

Many issues need to be discussed within the framework of a new outlook on priorities for the operation of the Savannah River Basin Lakes. The drought emergency as was recently experienced can be an opportunity to improve management options so that renewed crises with water availability for all purposes do not occur.

REFERENCES

1. "Savannah River Basin Drought Contingency Plan," US Army Corps of Engineers, Savannah District (March, 1989).
2. Dillman, B. L., and J. M. Stepp, Eds. *The Future of the Savannah River, Symposium Proceedings.* (Clemson SC: Water Resources Research Institute, June, 1976).
3. Abernathy, A. R. "Mercury Mobilization and Biomagnification Resulting from the Filling of a Piedmont Reservoir," Report No. 119 (Clemson, SC: Water Resource Research Institute, September, 1985).
4. Elzerman, A. W. "PCBs and Related Compounds in Lake Hartwell," Report No. 131 (Clemson, SC: Water Resources Research Institute, April, 1988).
5. Tisue, T., and T. Hsiung. "Manganese Dynamics in the R. B. Russell Impoundment," Report No. 127 (Clemson, SC: Water Resources Research Institute, June, 1987).
6. Cherry, R. District Chief. South Carolina U.S. Geological Survey, Columbia. Personal communication (1989).
7. Badr, A. W. South Carolina Water Resources Commission, Columbia. Personal communication (1989).

8. Hughes, W. B., M. S. Crouch, and D. Park. "Hydrogeology and Saltwater Contamination of the Floridan Aquifer in Beaufort and Jasper Counties, South Carolina," Report No. 158 (Columbia, SC: Water Resources Commission, 1989).

CHAPTER 14

The Savannah River System as a Microcosm of World Problems: Instructions to Conference Participants

Kenneth L. Dickson

INTRODUCTION

I have been asked by John Cairns, Jr. and Todd Crawford to develop some instructions or guidelines to assist the working groups as they develop consensus solutions/recommendations for the three case studies being addressed by the conference:

- Management of the Savannah River
- Endangered Species Protection — The Wood Stork
- Long-Term Management of the Savannah River Site Lands

The conference participants are to consider and apply the principles of integrated environmental management as solutions and recommendations are developed. A number of distinguished speakers have discussed various facets of integrated environmental management. These presentations have provided a common background and stimulated thoughts about the application of integrated environmental management to the case studies. Likewise, the presentations on the three case study topics have provided background information and have identified candidate problems and issues to be addressed during the conference.

OVERVIEW OF PRESENTATIONS

The purpose of this conference is to identify approaches for managing complex environmental problems and issues. The goal of the conference is to impact how complex environmental problems and issues are resolved in the United States and the world. The basic premise of the conference is that this can be done through the use of integrated environmental management (IEM). There are also some specific goals of the conference that relate specifically to the Savannah River Site. We have been asked to provide some ideas and guidance on potential uses for the Savannah River Site after the nuclear production facilities are decommissioned. Another goal of the conference is to stimulate IEM thinking about the current and emerging uses of the Savannah River and its associated basin. A final goal is for the conference participants to place biodiversity and endangered species issues in perspective and identify alternatives to address problems. Implicit in the conference purpose and goals is the development of a better understanding of the concept of IEM and its application to solving complex environmental problems.

What is *integrated environmental management*? Based on the presentations, there does not appear to be a clear-cut definition. John Cairns, in his presentation (Chapter 2), courageously offered a definition:

"Proactive or preventive measures that maintain the environment in good condition for long-range sustainable use."

Is this the "correct" definition? Are there other definitions that better communicate the concept of integrated environmental management? Conference participants are challenged to develop a consensus definition of integrated environmental management.

Based on the presentations during the symposium, it is evident that there are many complex environmental problems, such as deforestation, loss of biodiversity, sea level rise, ozone depletion, climate change, river basin use conflicts, etc., that can be addressed using IEM. However, many barriers prevent its use. Examples of barriers identified by speakers include the following:

- Focus on reductionism rather than integration
- Time constraints
- Turf battles between agencies
- Frontier environmental ethic
- High uncertainty of outcome of IEM
- Fear that management authorities will abuse power
- Changes in lifestyle required are strongly resisted by some

Conference participants are encouraged to consider these barriers during their deliberations and to identify ways to overcome these barriers.

The conference presentations identified a number of promising approaches for addressing/managing complex environmental problems and issues. Approaches identified during the conference include:

- Integrated resource management
- Systems analysis
- Decision analysis
- Geographic information systems (GIS)
- Risk assessment
- Expert systems

Although a number of potentially useful techniques/approaches exist that can be used to address complex environmental issues, it is highly improbable that any one approach can be used to address all problems and issues. The presentations and discussions during the conference suggest that the state of the science in IEM is ready for implementation. Working groups are encouraged to be creative in the uses of the concepts presented during the conference as they address their charges.

CHARGE TO THE WORKING GROUPS

It will be the job of each working group to:

- Develop an answer(s) to the principal question posed by the conference conveners.
- Identify the illustrative questions that must be addressed in developing a solution.
- Organize the illustrative questions into logical groupings.
- Develop a short (one-page maximum) exposition of each illustrative question.
- Identify solutions/recommendations for each illustrative question.
- Prepare a short narrative explaining the solutions/recommendations.
- Identify research needs.
- Produce, by the end of the conference, a rough draft of a chapter to be included in the book.

Meeting this charge will require a focused effort on the part of all conference participants. Excellent working group chairpersons will provide leadership and keep all to the tasks. However, since only slightly over one day remains to accomplish the job, several important considerations based on past conference experience should be helpful guides.

- Time does not allow each participant to tell "war stories." Each will have to resist the temptation to talk about previous experiences on interesting but vaguely related topics. Discussions will have to be focused and directed to the task. Unlike many conferences where a day or two is set aside for everyone to establish his territory, in this conference, the groups must get right to work!
- A brainstorming session should be used to identify illustative questions associated with each topic. This activity should be freewheeling. At this stage, no problem/issue

should be considered too trivial for listing. Listing of ideas on a flip chart is a good way to capture ideas. Illustrative questions considered should include those posed by the conference conveners, those identified by the presenters of the case studies, and those emanating from the discussions in the working groups.
- Organizing the illustrative questions to be addressed into logical thematic groupings will make it easier to develop the transitional linkages that make the written presentation read well.
- Members of the working groups should be given writing assignments to develop draft write-ups addressing the identified questions.
- All substantive input to the discussion should be generated at the conference. This means that the chairperson of each working group should have the verbage in hand at the end of the conference that will be needed to develop a draft of the written results.

CHAPTER 15

Management of the Savannah River

Paul Zielinski, Bernd Kahn*, B. Badr, P. Cumbie, J. Dozier,
J. Gordon, R. Lanier, M. Parrott, R. Patrick, D. Sheer,
R. Vannote, and N. Weatherup

INTRODUCTION

The working group on the management of the Savannah River agree that application of the principles of integrated environmental management (IEM), presented so eloquently earlier in the conference,[1] would be highly beneficial for a system as complex as the Savannah River Basin (see Figure 1). Not only are the many aspects of land use, water withdrawals, waste discharges, water quality flow rates, and environmental protection regulated by various state and federal agencies, but the objectives to be attained by these controls often conflict. It is envisioned that a more integrated management system would have the necessary data bases and mechanisms for conflict resolution to balance protection and development functions more effectively than is now possible.

Working group members recognize that pressure for applying this approach is far less intense on the Savannah River than at a number of river basins that are more highly industrialized and densely populated, with the associated water pollution and competition for water use. The group noted, however, that in view of the anticipated rapid population growth and economic development, this region in Georgia and

---

* Comments of all members of the working group were considered; however, this discussion was prepared by this designated person.

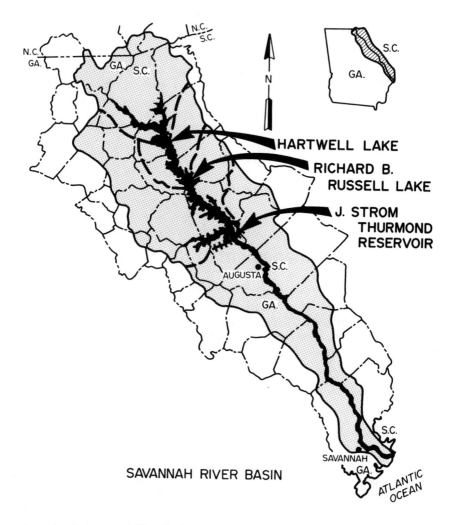

Figure 1. The Savannah River Basin.

South Carolina would benefit from having in place an effective management system. The system could be particularly beneficial in meshing the management activities of the sovereign entities — the federal and state governments — that today are separated functionally and by regulations in controlling one body of water, indistinguishable by state allegiance or federal primacy.

The management plan must be capable of integrating all the major functions of the river basin, both current and anticipated. To summarize the more detailed description,[2] three large dams on the Savannah River and three smaller dams on its tributaries produce hydroelectric power and control flow. At its mouth is the port of Savannah. The river provides cooling water for power plants, industrial complexes located mainly near Augusta and Savannah, and the large federal nuclear materials production complex at the Savannah River Site. These facilities, as well as a few municipalities, also use the river for water supply and as receiving waters for sewage and other waste effluents. Gold mining operations in the South Carolina mountains use water from tributaries. The river supports commercial and recreational fishing and recreational use of the reservoirs for boating and shoreline cottages. The river supplies water for and receives runoff from farms and contributes to silviculture in some flood plains. It is a habitat for biota, notably fish and wildlife, in the river, in adjoining marshes and flood plains, and on its banks. Its water is interconnected with nearby groundwater.

The availability of water and pleasures of the area should attract more residents, commerce, industry, and agriculture in the near future, and this growth will impose additional burdens on the river. Electric utilities envision additional power plants with cooling water requirements located at the reservoirs. Local and state agencies must consider the river for additional water supply. It is notable that water quality protection agencies in the two states have already assigned the entire waste effluent dilution potential considered available under current regulations to existing municipal and industrial users. Savannah envisions port expansion, supported by periodically dredged channels in the river. The United States Army Corps of Engineers, as operator of the three dams, must anticipate requests for modifying water release patterns to meet concerns for drought amelioration, recreational uses of impoundments, and biota habitat downstream, in addition to authorized operation for flood control, navigation, power production, and silt control.

The working group recommends the stepwise approach summarized in Table 1 for developing IEM of the river basin. This complex effort should be begun by examining each objective by itself, followed by an attempt to synthesize these objectives, and completed by selecting a management system. The initiation of this program will have to be the decision of the main participants in controlling the basin — Georgia, South Carolina, and the federal government.

According to this plan, each management objective will be developed by a task group of specialists in the respective field. The obvious topics are listed in Table 2, and others may well surface during the work of the task groups. Each objective will require information collection to provide a factual basis, calculational models to predict conditions under various circumstances, and plans for optimum development. Each task group must identify actual and potential problems, conflicts, limiting conditions, regulatory framework, supporting and opposing groups, benefits, costs, and risks.

Several approaches recommended by speakers at this conference are generally applicable and should be considered by all task groups. These include the geographical information system (GIS) for accumulating a sound data base,[3] a systems approach (SA) that quantifies benefits and costs in the form of calculational models,[4] and decision analysis (DA) that draws on a variety of fields and approaches to assist in developing a program acceptable to the interested parties.[5]

Water supply can be used as an example of applying a management plan.

Table 1. Developing and Implementing an Integrated Environmental Plan to Manage the Savannah River Basin

1. Identification of objectives
2. Examination of alternative approaches for attaining each objective
3. Consideration of mutualities or conflicts among objectives, emphasizing public involvement
4. Development of plan based on balancing multiple objectives and negotiated compromises
5. Implementation of plan by Georgia, South Carolina, and the federal government

Table 2. Objectives for Savannah River Basin Management

1. Preservation of natural system/aesthetics
2. Water quality control
3. Water supply availability for utilization, cooling, and dilution
4. Hydroelectric power
5. Navigation
6. Flood control
7. Drought mitigation
8. Industrial and commercial development
9. Agriculture and silviculture development
10. Fish and wildlife conservation
11. Recreation
12. Land use optimization
13. Water availability for groundwater recharge
14. Wetlands maintenance

Potential demand should be considered at various locations along the river by municipalities and industry as a function of time, and this demand should be matched to potential water availability as affected by climate, controlled river flow, and competing uses for both surface and groundwater. To the extent necessary, water use may then by controlled by pricing, supply regionalization, conservation practices, partial reuse, or reducing losses. Reservoirs and supplemental sources can be used for periods of low rainfall. Guaranteeing a steady supply from the river, however, may interfere with reservoir management plans for power production, recreational uses, flood control, or wetlands maintenance. Natural droughts and man-made pollution may interfere with delivery of a suitable water supply, and the quality of the returned wastewater may adversely affect biota in the river. Each set of conditions will impose a limit on population and industrial users.

An approach to managing fisheries and wildlife in the river basin is discussed by R. S. Vannote in the appendix. Distinct ecological habitat must be considered with regard to optimum management, limitations and controls, exising problems, and interactions with other uses of the basin. It is clear that every other cited objective interacts in some manner with maintaining habitats, and that the extent of support for this objective has to be balanced with support for every other objective.

Attainment of multiple objectives is conditioned by these limits set by the plans for individual objectives. Maximum utilization of the river for one objective may be devastating for another, benefits, be they ecological or financial, for one objective in the short term may have disadvantages over the long run for another. Favorable objectives for one group may appear destructive to another with different priorities.

Dedication to achieving optimum benefit from the river basin, technical ingenuity in combining the best of the various plans, and skill in achieving compromises will be needed when the planning committee takes the products of its individual task groups to prepare the integrated management plan. The primary recommendation by the working group is that committee membership be inclusive to the fullest possible extent. All groups with significant interest in the future of the basin need to be represented: elected officials and administrators for the states and their local subdivisions, federal officials, impacted populations and industries, and public interest groups. Furthermore, the deliberations of the committee must be open to an extensive public so that the recommended plans will have the benefit of the widest possible scrutiny and debate. Every point of view should have been considered fairly before a plan so complex and wide ranging is prepared.

The following issues must be addressed and incorporated into the management system:

1. Reasonable allocation of the water supply and the assimilative capacity of the river and tributaries among Georgia, South Carolina, and federal needs.
2. Establishment of water quality regulatory standards that the three entities find acceptable.
3. Specification of indicators that minimum acceptable levels for water quality, habitat, recharge, etc., are being met.
4. Consistent approach to wildlife/fisheries regulations and habitat maintenance.
5. Optimized water availabilites for in-stream and off-stream use, including habitat considerations, by controlling flows and reservoir impoundment levels.
6. Land use management and controls.

The main components of the integrated management plan will be the set of combined and interrelated objectives, supporting factual documentation for the objectives and quantitative evaluation of the benefits and costs compared to the alternatives that were rejected. Preparation will have to be assigned to technical task groups that will make full use of the previously applied GIS, the SA calculational models, and presentation of alternatives. An important outcome of this effort will be identification of questionable assumptions and predictive uncertainties that will have to be resolved by specific studies or measurements.

Equally important are recommendations of the structure of the system for managing the program and guidelines for operation. Precedents exist for a wide range of options. Such alternatives include assignment of lead responsibilities to various existing federal and state agencies, designation of an advisory group to guide cooperation of various agencies, assurance of consistent responses by the various agencies through an interstate compact, or establishing a basin authority either working through existing entities or with full management powers and staffing. Although the working group members believe that focusing responsibility is vital to effective management, the group realizes that it would not be easy for the various federal and state agencies to surrender major responsibilities for controlling the river and its shores.

It is important that the management organization have the capability of monitoring the outcome of its control of the river basin with regard to factors such as consistency of water flows and quality with modeling predictions, well being of biological indicators, economic value of the basin, and satisfactory utilization of water for various applications. At the same time, management must also accumulate information based on measurements, surveys, and literature reviews that indicate trends and tendencies due to new findings and ideas, and macroscopic changes in climate, population trends, and economic well being. Such monitoring and information gathering should be in part internal to the management, but can be supplemented externally.

Prescribing an IEM program for a river basin, based on thorough informational input, with conflicting objectives resolved through wise mediation and implemented by a strong and fair manager is relatively easy. Making it work in a world of imperfect knowledge, bitter diversity concerning objectives, and reluctance to share responsibility among agencies will be a difficult task. That a general consensus was reached in this group concerning the approach is a good beginning; however, the group strongly recommends pursuing this process to manage the river basin even though it is not now in desperate need of restoration.

APPENDIX

Fisheries and Wildlife: by Robin L. Vannote

Wetland River Habitats

This analysis primarily considers fish and wildlife resources associated with wetlands and stream riverine systems, exclusive of upland agriculture and forest lands, although they are critically important in wildlife management and plans. Fisheries and wildlife management objectives to consider are

- Identification of methods for optimizing management
- Determination of present controls or limitations

Six ecological habitats can be identified for incorporation into the Savannah River environmental management plan:

1. Coastal and tidal river estuarine systems

2. River flood plain swamp-forest: swamp system
3. River system (main stem below dam)
4. Coastal tributary systems, including their associated wetlands
5. Upland tributary system
6. Reservoir systems

In brief, the objectives are to maintain, enhance, and replace lost habitats. In fisheries and wildlife, habitat is the crucial element in attaining management goals.

Matrix to Identify Conflicts, Barriers, and Reinforcement Areas

	Interactions with other resource uses		
	Integration with fish and wildlife		
	Enhance	Neutral	Adverse patented
Management objectives	+	0	-
Water supply use (domestic)			-
Water quality	+		
Water quantity		±	
Fisheries and wildlife			
Agriculture			-
Forestry		±	
Mineral development			-
Groundwater recharge	±		
Recreation		±	
Navigation			-
Maximize land use value			-
Industrial/commercial			-
Aesthetics	+		
Flood control			-

To protect the six identified fish and wildlife systems within the Savannah River, the following construct is outlined.

1. Coastal and Tidal River Estuarine Systems

A hydrological water management plan must be developed that identifies the optimal water levels (elevation) and volume of circulation needed to achieve seasonal requirements. Boundaries should be identified for plant/animal communities.

How the present water allocation from storage reservoir conflicts or aids estuarine management goals should be identified. The volume of river discharge needed to attract anadromous fish (e.g., striped bass) into the harbor and river and if the water quality of the harbor during periods of high local runoff is adequate to assure safe passage of fish should be determined. What quality of water is required to meet estuarine management objectives?

2. River Flood Plain Swamp-Forest: Swamp System

On large river systems such as the Savannah, the adjoining wetlands must be considered an integral part of the entire system. Flood plains on natural, unregulated rivers are inundated about once per year, and many wetlands and riverine species have evolved and become dependent upon seasonal flooding. Storing water by flood skimming interrupts this critical activity, leading to loss of habitat species and near surface groundwater recharge.

Questions that require research include the minimum and optimum flood frequency schedule and maximum stage necessary to maintain wetlands. Time periods when exchange (nutrient, fish, invertebrates) between wetlands and riverine systems is critical must be identified, and river stages necessary to ensure this exchange must be determined. The relative importance of winter-spring floods versus wetlands inundation during other seasons, e.g., late summer or fall, must also be determined. Is water quality a factor in allocating water for flooding wetlands?

3. River System (Main Stem Below Dam)

The main stem of the Savannah River extends from the fall line (J. Strom Thurmond Dam) to the seasonal salt water line near Savannah, GA. Maintaining and enhancing fisheries and wildlife values are the principal goals. The research requirements are to identify the hydrologic regime necessary to achieve management goals, to determine optimal and minimal seasonal discharges, and to review water quality conditions throughout the river reach to assure that allocated release from storage reservoirs is sufficient to meet fisheries and wildlife requirements.

Flow volume of the Savannah River must be sufficient to attract anadromous fish into the harbor and up river. What is the timing (season) and volume of discharge necessary to provide attraction and assure safe passage in the event of deteriorating water quality during periods of significant local surface runoff? Many resident fish populations require seasonal access to river flood plains and wetlands for spawning and nursery activities. Late winter and early spring is the usual period when fish movement occurs, and success is dependent upon elevated river stage to provide passage through breaks in the natural levee and drainage tributary streams. River stage must also be sufficient to inundate the wetlands and provide sufficient habitat for fish fry development and survival.

Additional studies are required to determine the minimum flows necessary to maintain river habitats in summer and autumn, a period of normal low discharge. It is necessary to identify river habitat critical for producing trophic resources for fish population and essential cover habitat during the low flow seasons. Water allocation must be based on seasonal needs for maintaining fish and wildlife resources.

The Savannah River presently has a resident population of a rare and perhaps endangered species of fish: shortnosed paddlefish sturgeons. Factors regulating life history parameters must be identified and incorporated into regional environmental planning.

4. Coastal Tributary Systems, Including Their Associated Wetlands

Under IEM, will changes in storage and allocation of water from main river reservoirs shift the increased water demand to tributary streams and groundwater? Such a shift may lead to construction of water supply reservoirs on tributaries, leading to loss of wetlands habitats and stream communities. Increased groundwater withdrawal may diminish summer and autumn flow levels in headwater streams with deterioration of stream and wetland habitats. Diminished base flow level may contribute to lowering water quality and habitat degradation.

Tributary flood plains and wetlands frequently are inundated because their outlet is flooded by high stage on the Savannah River. Blocked outlets back tributary water up the channel, flooding its banks. This flooding cycle may occur without local storm events and is an important cycle in maintaining tributary wetlands and inducing bank storage of water.

Research is required to document the flood frequency necessary to maintain wetlands along coastal plain tributary streams. Do the same time frequency or magnitude of floods that apply to the riverine system apply to the tributary-wetlands ecosystem? Are biotic processes in tributary systems essential for maintaining riverine systems? What are the major resource uses within the basin that impact tributary-wetlands systems and can these be ranked?

5. Upland Tributary System

In the Savannah River Basin, the upland streams are located in the Piedmont and mountain providences. The major fisheries and wildlife goals are protection and enhancement of existing resources. The major issues confronting these goals are water diversion and headwater impoundment for water supply to meet development and expanding industrial needs; definition and documentation of the importance of headwater streams and rivers to recreation, fisheries, and wildlife; and improvement of water quality in Piedmont streams with respect to sediment input, transport and storage, nutrient loading, and toxic chemicals.

The major issues regarding preservation, enhancement, and restoration of stream and wetland habitats largely focus on water allocation, water quality, and land use regulation factors. The challenge is to devise a central compromise position that assures preservation of habitats in a sustainable productive state to meet the requirements of an intact ecosystem. An agenda must be developed to integrate fisheries and wildlife objectives with the compromised position of other agencies and communities. Barriers to meeting fish and wildlife goals require definition in quantitative terms.

6. Reservoir Systems

The three main channel reservoirs are high centers of fisheries, wildlife, and water recreational activities. Presently, water level management is optimized for flood control, power production, and maintaining a satisfactory water elevation for summer boating recreation. The fisheries and wildlife management goals should include a water budget (seasonal elevations) that would optimize reproduction growth and recruitment of sport and forage species. Waterfowl management objectives are also controlled, in part, by water allocations, and these needs must be identified and weighed against other allocation demands.

SUMMARY

The management objectives for Savannah River Basin are to maintain, enhance, and restore river and wetland habitats. The goal is to support a natural and diverse flora and fauna. Important issues are

- Allocation of water from storage reservoirs in maintaining riverine and wetland habitats.
- Determination of how people within the basin perceive the value of the fisheries, wildlife, and wetlands and if this is supported by use.
- Determination of how important riparian wetlands are modifying or altering the biogeochemistry of groundwater moving from upland recharge areas through bottomland riparian zones and into stream channels.

Is maintaining a certain *quality* of water an important objective for the Savannah River Basin?

Advantages	Disadvantages
I. Prevention of salt water intrusion in lower Savannah River	1. May diminish amount of impoundment water for users, such as recreation, and maintaining land value and aesthetics
2. Maintaining water quality by sufficient water for diluting discharges from industry, city sewage, etc.	2. Reduced hydroelectric generation
3. Maintain wildlife and endangered species habitats	
4. Maintain water quality for drinking purposes	

Some ways to increase water quality available to users that do not require increased releases are

1. More efficient irrigation systems, use of night irrigation, use of sewage water.
2. Domestic conservation, e.g., low-flush toilets, use of drought-resistant landscaping.
3. Industry conservation, e.g., recycling, processing water, using sewage water.

Is *groundwater recharging* an important objective for the Savannah River Basin?

Advantages
1. More dependable and stable water supply (less fluctuations)
2. Filtering of pollutants from system, i.e., better water quality
3. Enhance wildlife and fisheries management
4. May protect habitats of endangered species
5. Protection of recharge areas increases areas available for recreational hunting and fishing
6. Protected areas are likely to be more aesthetically pleasing

Disadvantages
1. Recharge areas must be protected and cannot be developed either privately or commercially; agriculture would also be limited
2. Public must be "sold" or educated as to the values of recharge areas
3. Recharge areas need to be protected and have a protected river corridor (50—100 ft floodline) established; this restricts development

How are recharge areas identified (this would be done by interstate cooperation with air surveys)? Groundwater recharge can be accomplished by

Manipulation of water discharges from the Corps of Engineers impoundment:

Advantages
1. Water is already stored in dams
2. No new structures need to be built, thus minimum expenses involved

Disadvantages
1. Possible opposition from impoundment residents, especially during drought conditions
2. Possible fluctuation in hydroelectric generation

Off-river storage:

Advantages	Disadvantages
1. Storage water can be collected at optium times with little impact on impoundments	1. Building these structures removes land for alternative uses (agriculture, wildlife, etc.)
2. Can be used to supplement river flow	2. Expenses associated with construction

Water Quality

Water quality objectives for the Savannah River are set forth in water quality standards established by the South Carolina Department of Health and Environmental Control (SC DHEC) and by Department of Natural Resources (DNR), with oversight by the United States Environmental Protection Agency. In order to achieve these objectives, the water quality of the river has been managed by the SC DHEC and by DNR by regulation of grant source (municipal and industrial wastewater discharges) through the NPDES permit program. This program establishes maximum effluent loadings that can be released into the river system by each discharger under a set of perdetermined critical conditions. Monitoring by both states indicates that water quality in the river from its headwaters to Clyo has been generally good. Water quality in Savannah Harbor has improved drastically over the past 15 years; studies are ongoing to ascertain whether the quality in the harbor is sufficient to protect critical fisheries.

Quality Control

Quality control of the Savannah River has been strengthened by the adoption of nonpoint source management programs by both states. These controls depend on operational minimum releases from Lake Thurmond, which have recently been questioned in light of the implementation of the Corps drought management plan. In addition, water quality modeling and actual monitoring data indicate that little, if any, additional assimilative capacity remains in the river system from Augusta to the ocean. Therefore, releases from the lake must be managed to optimize the water quality of the Savannah River to achieve a balance among competing users in the system. Future development of the Basin can be accomplished:

1. By introducing no additional pollutant loading (i.e., reuse, land application, or the other no discharge alternatives)
2. By reallocating existing loading from permitted dischargers to new dischargers
3. By modifying release schedules from Lake Thurmond to provide additional assimilative capacity during summer months

REFERENCES

1. Cairns, J., Jr. "The Need for Integrated Environmental Systems Management," this volume.
2. Zielinski, P. B. "Impacts of Management Decisions on Environmental Issues of the Savannah River," this volume.
3. McHarg, I. "Land Use Planning for Multiple Uses," paper presented at Integrating Environmental Management Symposium, Aiken, SC, September 26-28, 1989.
4. O'Neill, R. V. "The Systems Approach to Environmental Assessment," this volume.
5. Kramer, R. A. "Decision Analysis as a Tool in Integrated Environmental Management Issues," this volume.

CHAPTER **16**

Endangered Species Protection — The Wood Stork

**William D. McCort, Nora A. Murdock,* I. L. Brisbane,
F. B. Christenson, M. C. Coulter, J. R. Jansen, R. A. Kramer,
S. Loeb, H. E. Mackey, T. M. Murphy, and P. Stengle**

At present, there are 535 species of plants and animals that are federally listed as endangered or threatened, and the number is increasing by approximately 50 species per year. The rate at which these species are being added to the federal list is reflective only of time, manpower, and funding constraints, not of the number of species that truly merit the designation. The latter number is believed to be much larger since there are currently about 4000 species that are "candidates" for federal listing as endangered or threatened.

WHY SHOULD WE BOTHER ABOUT SAVING THESE VANISHING SPECIES? WHAT GOOD ARE THEY?

The following examples illustrate the economic importance of natural diversity:

- Half of the prescription drugs used today came originally from plant or animal material. The commercial value of drugs (nonprescription and pharmaceuticals) in developed countries in 1980 was approximately $40 billion/year.
- Children suffering from leukemia 30 years ago had approximately a 20% chance

---

* Comments of all members of the working group were considered; however, this discussion was prepared by this designated person.

of survival. Today, thanks directly to a relatively obscure plant known as the rosy periwinkle, survival chances of this disease's victims have increased to 80%. The commercial sales of materials from this one plant worldwide now exceed $100 million/year.
- Over 3 million Americans with high blood pressure today owe their lives to a species of foxglove (and digoxin, the drug discovered in this plant's tissues).
- One class of compounds, the alkaloids, which are almost exclusively derived from plants, are extraordinarily diverse in structure and biological activity. Thousands have already been chemically characterized and hundreds of these were found to have practical uses for human beings, including analgesics, narcotics, anesthetics, insecticides, parasiticides, antimalarial and antiamoebic agents, muscle relaxants, respiratory stimulants, gout depressants, and anticancer drugs, to name a few. In addition, 5000 flowering plants have been analyzed for the presence of alkaloids, yet this is only 5% of the 250,000 species of flowering plants in existence. There are even more species of animal life, most of which have not yet been screened for commercially valuable compounds.
- Agricultural values of biodiversity are equally impressive. The wild relatives of commercial crops serve as genetic reservoirs, containing genes for disease resistance no longer possessed by commercial strains. There are also less expensive and less damaging alternatives to chemical pesticides found in native plant and animal material. (An illustration of how unique genetic material found in wild plants can have direct and significant economic impacts on commercial agriculture is exemplified in the recent discovery of a wild perennial corn found growing in a remote montane forest in southwestern Mexico. The discovery and preservation of these seemingly insignificant plants may well result in grain farmers saving billions of dollars per year in the cost of preparing and replanting fields. Erosion would be significantly reduced, thus maintaining the value of the land and improving water quality in areas downstream.)
- Industrial examples of the value of biodiversity are equally impressive. Products now being found in wild plants and animals are proving to be very effective (and in some cases, superior) substitutes for petroleum products. Development of these alternate sources is clearly advisable since petroleum deposits are finite and economic independence from the OPEC nations would thus be furthered. (Example: each American citizen uses 500 lb of petrochemical products per year. In 1973, this consumption cost approximately 3¢/lb; the cost has now increased to 22¢/lb. In contrast, vegetable fats and oils cost 1¢/lb in 1973; today, they are still less than 2¢/lb.)

Other justifications for preserving endangered species and thereby contributing to the preservation of overall biological diversity include arguments for aesthetics — do humans really want to live in a world where there are no more bald eagles, polar bears, tropical orchids, etc.?

There is also the ethical argument. Even if the decision is made to "let some of these species go," who is wise enough to choose which ones? Before penicillin was discovered and saved millions of human lives, how many, even as biologists, would have fought to save this species of mold?

Perhaps the most important value of endangered species is their role as sensitive indicators of overall environmental health and integrity. A commonly used analogy is that of individual bricks in a large building, being removed one by one. The removal of one, or even several, may not be noticed immediately, but sooner or later overall structural integrity will be affected, and the removal of one more brick will cause the building to collapse. To paraphrase one of Aldo Leopold's more famous quotes: the first step in intelligent tinkering is to remember to save all the pieces.

These ideas can be focused on the local example of the wood stork and L-Reactor. In 1983 and 1984, a controversy arose over the restart of an old nuclear reactor on the Savannah River Site vs. the endangered wood stork. The Endangered Species Act of 1973 contains two major mandates for federal agencies: (1) each federal agency is directed to ensure that any action it authorizes, funds, or carries out is not likely to jeopardize the continued existence of any endangered or threatened species or result in the destruction or adverse modification of critical habitat for such species; and (2) all federal agencies are directed to utilize their authorities to carry out programs for the conservation of endangered and threatened species in furtherance of the purposes of the Act. In the 1930s, there were 20,000 nesting pairs of wood storks in the United States; at present, there are approximately 6000 pairs. Those storks surviving in traditional nesting habitat are failing repeatedly to reproduce (successfully fledge young). What is the issue here? Just a stork? One bird? The picture becomes clearer when the reasons for the decline of the species are considered. The wood stork is the focus or indicator species in this case, but there are currently only 1% of the wading birds (of all species) in the Everglades that once existed there. The cause for these declines is primarily massive disruption of hydrological systems in South Florida, which is now beginning and will continue to affect drinking water and agricultural/industrial water supplies for millions of people living there, as well as for the wild species of fauna and flora sharing that overburdened habitat.

This points to the original question: "What magnitude and type of effort should be expended on protecting endangered species?"

The magnitude should be scaled to the relative severity of the threat to the system, and a commensurate amount of resources should be allocated to resolve the problem. For the wood stork, this magnitude is going to be considerable because the situation has graphically deteriorated. What type of effort is required? The Endangered Species Act, although it focuses on individual species for enforcement and implementation, is a means for conserving the ecosystems upon which endangered species depend. The idea of integrated management ties in closely with ecosystem management. The wood stork example goes from species-specific (wood stork) problem solving to group (wading birds) management to community management (foraging and nesting habitat such as that of the Savannah River Swamp System) to preservation of biodiversity.

The ultimate goal of the Endangered Species Program is recovery. For some species, such as the brown pelican and the American alligator, the problems are relatively straightforward and easily remedied. The principal problem for the pelican was organochlorine pesticides; once the use of these was discontinued, the bird began to recover. For the alligator, all that was required was protection from unregulated hunting. Some problems are not so easily alleviated. The latter cases will require research followed by applied management based on research results. However, proactive approaches to problems are much more effective and much less expensive than reactive approaches.

The fundamental reason for protecting endangered species and preserving biological diversity is the need to maintain ecosystem stability. The problem of endangered species management is monstrous already and growing in magnitude every year. The goal of preserving overall biodiversity is even larger. Therefore, to have any hope of dealing successfully with these problems, management approaches must be integrated across agencies, disciplines, and levels of biological organization. Exclusive specializations and petty turf battles are luxuries that can no longer be maintained.

Participants at this conference have discussed management of the Savannah River Basin, the Savannah River Site itself, and endangered species, but there is really just one issue here — management of the global ecosystem — the ultimate integrated system, of which humans are inextricably a part. Although it has become a cliche in many circles, the following statement bears repeating: the fate of endangered species is ultimately our fate.

REFERENCES

Specific examples and statistics in this discussion are taken from or are influenced by these publications:

The Nature Conservancy News.
Koopowitz, H., and H. Kaye. *Plant Extinction: A Global Crisis* (Washington, D.C.: Stone Wall Press, 1983).
Myers, N. *Sinking Ark: A New Look at the Problem of Disappearing Species* (Oxford: Pergamon Press, 1979).
Wilson, E. O. *Biodiversity* (Washington, D.C.: National Academy Press, 1988).

CHAPTER 17

Long-Term Management of the Savannah River Site Lands

F. Ward Whicker*, Jerry J. Cohen, S. Bloomfield, K. L. Dickson,
E. C. Goodson, J. G. Irwin, M. N. Maxey, R. V. O'Neill,
J. F. Proctor, L. E. Rodgers, and S. R. Wright

INTRODUCTION

The broad question discussed by this working group was *"What should the long-term land management strategy be for the Savannah River Site (SRS)?"* This question is particularly relevant at this time because the future of the nuclear materials production mission is clouded and because the Department of Energy (DOE) is changing its top priority from nuclear materials production to health, safety, and environment. The general goal of the land management strategy is to optimize and balance national needs with the protection of human health and enhancement of environmental quality.

A number of potential long-term uses for the site were discussed and among those, three were unanimously endorsed by the working group:

- Ecological preserve
- Environmental research park
- Waste disposal site

---

*Comments of all members of the working group were considered; however, this discussion was prepared by this designated person.

A fourth potential use was discussed, namely that of a nuclear reactor site. Under this category, at least three options are possible, including:

- Continuation of weapons material production
- Nuclear electrical power production
- Decommissioning of nuclear facilities

The working group made no attempt to choose among the latter alternatives. However, the particular choice may well affect certain land use policies, such as public use of the site, cleanup standards, and schedule, etc.

ECOLOGICAL PRESERVE

It is important to clarify briefly the first three uses identified because their titles alone inadequately describe them. The ecological preserve is meant here to denote a relatively pristine area that is managed in a way that offers a high degree of protection from damage or contamination. The preserve would be expected to remain undisturbed and to change primarily by natural forces. However, it is not intended that justifiable management actions be excluded in the area. For example, controlled burns and pest control would not necessarily be excluded indefinitely, nor would low-impact research be precluded.

ENVIRONMENTAL RESEARCH PARK

The environmental research park would denote a broad zone within which basic and applied research would be encouraged. The size and diversity of the SRS is compatible with a wide range of research activities. Basic work in fields such as hydrology, geology, atmospheric science, and ecology would be feasible on the site. The site would be particularly and uniquely suited for studies in landscape ecology, global change, biodiversity, ecotoxicology, and radioecology. Applied research on topics such as clean-up technology, site remediation, contaminant transport, etc., could also be conducted more readily at the SRS than in any other ecologically comparable area.

WASTE DISPOSAL SITE

With respect to waste disposal, there was a strong consensus that complete removal of existing chemical and radioactive wastes at the SRS is not feasible, cost effective, or necessary. Indefinite storage of many waste forms, excluding reprocessed high-level radioactive wastes, is therefore favored. A strategy to minimize new wastes in the future is essential and, should the SRS be decommissioned, new wastes generated elsewhere should not be disposed of at the site because of unfavorable hydrogeology.

A number of issues are related to the long-term uses of the SRS, and some of these were addressed by the working group.

- Need for risk-based cleanup standards
- Regulatory concerns
- Lack of credibility
- Need for education and involvement of the public
- Need for technical expertise
- Ethical concerns
- Communication to achieve implementation

With respect to the need for risk-based standards, there was strong consensus that science and logic should largely drive the decision making process. Failure to use this approach will perpetuate many current problems, create new ones, and lead to large, unnecessary monetary costs. The working group also felt that many people accept too readily the illogical regulations that stem from ignorance or fear. When such regulations appear, they should be aggressively challenged.

A major problem across most of the DOE nuclear weapons complex is lack of public credibility. This problem also plagues the SRS. Irrespective of what the truth is, the public at large is not likely to believe it. No quick solution to this problem is evident. However, a long-term approach to this problem may exist. This approach includes better education and more complete involvement of the public. Public involvement in decision making, planning, etc., should be encouraged. Involvement of the SRS in public education, at all levels, will also be required.

Another problem for the SRS is the lack of technically trained personnel in certain fields. Those that relate to the issues raised by this working group include ecotoxicologists, radioecologists, environmental health physicists, waste management specialists, systems ecologists, etc. Many of the regulatory problems stem, in part, from a lack of technical parity among DOE, their contractors, and the regulatory agencies. More emphasis on relevant education will eventually remediate this problem, as well as the problem of credibility mentioned earlier.

A major concern of the working group was "will this exercise have any impact?" Clearly, unless the ideas from this conference are effectively communicated, they are not likely to be implemented. The ideas need to be communicated to regulators, the public, Congress, and the academic world.

CHAPTER 18

Future Needs

John Cairns, Jr.

> "If you make people think they think they will love you.
> If you really make them think, they will hate you."
> Don Marquis

INTRODUCTION

I will not attempt to summarize the entire book here because subcomponents have been summarized elsewhere. The genesis of these remarks is the talk given at a public meeting scheduled for the close of the conference at which various representatives of the news media and environmental organizations were present. If one accepts the need for a less fragmented, more integrated approach to environmental management, the question is where do we start? The answer is, of course, everywhere simultaneously. It is abundantly clear that the loss of topsoil, acid rain, increased greenhouse gases, population growth, destruction of forests, and the high rate of species extinction globally cannot continue indefinitely. Therefore, we must turn from exploiting the environment to maintaining the environment in a condition suitable for sustained use. This means integrating environmental management so that one use does not impair other beneficial nondegrading uses. Also, it means ensuring that the aggregate of all environmental uses permits maintaining natural systems in robust health for sustained, long-range beneficial use (i.e., continued ecosystem services).

While the science for making predictive models of effects of various courses of action upon the environment needs to be far more robust than it is now, undoubtedly the major determinant in the success of integrative environmental management (IEM) will be social and political will. In short, the general public and the global assemblage of nations must have a majority endorsing this approach. The general public and its representatives must have a clear overview of the problem. As Thomas Jefferson said: "I know of no safe depository of the ultimate powers of society than the people themselves, and if we think them not enlightened enough to exercise their control with a wholesome discretion, the remedy is not to take it from them, but to inform their discretion." I have little doubt that the approach will ultimately be embraced, but I fear that enormous irreparable environmental destruction will be required for the development of the necessary social and political will. Multiple catastrophies may have to occur on a global scale to motivate the politicians and the general public. Preventive action is desirable if it could be taken in time to mitigate or perhaps even avoid some of these catastrophies. There is some indication that the general public and some politicians are ahead of mainstream science in this regard.[1] However, history indicates that no substantive action will be taken until the crisis is unmistakable to even the most unobservant people.

In the meantime, however, much can be done to improve the present situation. Some future needs follow. These remarks are entirely my own and do not necessarily represent the thoughts of others in the conference or those of my co-editor, Dr. Todd V. Crawford. Undoubtedly, some (perhaps all) of the participants share some of my views and perhaps even one or two all of them. Nevertheless, I take full responsibility, as I should, for all of these remarks.

THE TYRANNY OF THE DISCIPLINES

The title for this section comes from the first sentence of "Part VIII — The Academic Organization" in a draft report entitled "The Case for Change" (Commonwealth of Virginia, Commission on The University of The 21st Century, 101 North 14th Street, Richmond, VA 23219). The second sentence of that section of the document states: "We understand this phrase to mean that the academic disciplines and departments that support them define acceptable methods of inquiry and what it means 'to know' something about ourselves and about the world." The report further states: "Discipline-based departments set the criteria by which research, scholarship, and teaching are evaluated and, as a result, how rewards are meted out to faculty members (promotion, tenure, salary increases, teaching schedules, research space, and so on). Membership in a discipline and the corresponding department, rather than in a particular college or university community, is the basis for many faculty members' professional lives."

The report states that the president of one Virginia university has observed the fact that much exciting teaching and research that is called "interdisciplinary" is really a mark of shame; the present disciplines are no longer adequate for what we know and the problems we must solve. Nevertheless, academic departments exert too much control in colleges and universities. As a result, the rewards for working outside the established boundaries are limited. For junior faculty, unprotected by tenure, the sanctions often are fatal. I hasten to add the caveat that I do not advocate abolishing the disciplines because they are very useful, although artificial, constructs that have improved communication among specialists. The tyranny is that they effectively insist that everyone should be a specialist, as noted by the report just cited. I have seen promising young faculty members with superb research and teaching evaluations denied promotion in a number of academic institutions because their activities did not fit a disciplinary bias.

Perhaps the "tyranny of the disciplines" exists because tenured faculty are justifiably insecure in this era of future shock. Complex multivariate problems cause feelings of insecurity in the most holistic research investigators with the broadest views. However, the nonconstructive defensive response is to force all members of a discipline to fit neatly into a particular niche where it is easier to keep abreast of what other occupants of the niche are doing and simultaneously prevent them from infringing on one's particular subspeciality or "turf." This is a form of resource partitioning and works well on a disciplinary "island," such as a department. However, funding agencies are increasingly determining which problems will be investigated, thereby removing a substantial portion of resource allocation from departmental or even university level decisions or authority. The departments may still exert control in determining (through tenure, promotions, space assignments, teaching loads, etc.) which subspecialities prosper (for example, molecular biologists have successfully excluded ecologists and taxonomists from some biology departments), but extramural funding agencies will ultimately determine the overall direction of research. As the world crises increase in intensity and frequency, almost certainly more research money will be diverted to firefighting on major problems transcending disciplinary boundaries. The discontinuity between the national and international resource allocation and the resource allocation within the department may grow greater and greater for those departments unwilling to adjust to the present dramatic trends. The institutions that flourish over a great many years will be those in which reductionists and integrative science co-exist amicably or, better yet, creatively, and those that decline precipitously will be those fighting the last war in which specialization was the only way to go. I used the Jefferson quote at the outset because ultimately the citizens themselves will determine how well the university fits future needs. Clearly the report of the Commission on the University of the 21st Century calls for fundamental changes in higher education definitely because of clearly identifiable faults within the present system.

There is a compelling need for integrative science (actually, the integration should go well beyond science to include the humanities, etc., so the words should be *use of integrative studies* in the assemblage of evidence or knowledge for making decisions on global or large scale problems). This requires more than an assemblage of specialists each going to the ultimate level of detail in his or her specialty. It requires a holistic view that does not prejudge the kinds of evidence necessary to make an effective decision. The use of decision analysis[2] will help determine both the kinds of information and the level of detail necessary to make an effective decision. This is the constructive way to reduce professional insecurity.

THE ACADEMIC RESPONSE TO THE TYRANNY OF THE DISCIPLINES

In the late 1960s and early 1970s, at least some academic institutions recognized that global problems (such as overpopulation, soil loss, pollution, etc.) did not fit neatly into the disciplinary compartments erected on campuses. The response was to create interdisciplinary centers following one of two general strategies:

1. The center was set up as an institute, department, division, or some other academic construct modeled after more traditional, existing units on campus. In this model, faculty were tenured within the unit and degrees granted, both undergraduate and graduate, as they are in traditional discipline-oriented administrative units.
2. The second alternative was to form a center with no tenured faculty and no degree granting authority that would draw participants from a variety of disciplines into teams for solving specific problems. To the extent that students were involved, usually in research only, they accompanied the faculty members from different disciplines. No degrees were granted except in the traditional disciplines.

Advantages and Disadvantages of Strategy 1

An advantage to strategy 1 is that talented young faculty with interdisciplinary interests are not penalized during the tenure and promotion process because they participate in interdisciplinary activities and are, therefore, "disloyal" to their home discipline. In addition, because degrees are granted within the interdisciplinary unit, students who take a series of interrelated courses do not spend time at the whims of an outside funding agency (which happens when only extramural funding supports interdisciplinary activities) but in the fulfillment of a degree. Therefore, from a teaching standpoint, the program is stable and predictable rather than being driven by extramural funding.

A disadvantage to strategy 1 is that no interdisciplinary department to my knowledge has an adequate array of talent to fit the needs of vast problems, such as the greenhouse effect or global warming, in-depth hazardous materials studies, etc. Typically, such teams consist of hard scientists, engineers, and occasionally an economist. The humanities, social sciences, etc. are generally left out, although they should not be. *The point is that if a university or college were truly a community of scholars, participants in an interdisciplinary team could be drawn from a larger talent pool than is possible from a single department.* The university is severely handicapping its capabilities for interdisciplinary work by limiting the pool from which interdisciplinary teams can be drawn to those departments or disciplines willing to tolerate "outside" activities. Even when this tolerance is present, tenure and promotion committees consisting of specialists within the discipline do not do a good job of judging the qualifications of candidates for promotion and tenure if they are interdisciplinary.

In addition, team members in a department get used to working with each other and deficiencies in particular types of experience can be overcome by being a perpetual "student." Sometimes the aggravation of getting discipline-oriented people to accept the fact that rituals and practices in other disciplines differ from their own is too time consuming and, therefore, the flexible mindset needed for interdisciplinary work may be more valuable than a particular specialization.

Advantages and Disadvantages of Strategy 2

Any activity on a university or college campus that cannot fully involve students is inappropriate. Forcing a student to meet the requirements of a discipline and, at the same time, those of an interdisciplinary activity is possible for the brightest students who can beat the system regardless of the rules, but is not always possible for the average student. Many problems requiring interdisciplinary teams are by necessity long-term efforts and the stability of a team assembled from departments where this is not a primary goal is not adequate for this task. Since no one is tenured in the interdisciplinary center, persons who do not fit interdisciplinary activities are quickly weeded out. Unfortunately, this process is a great strain on a team because one or two defaulters are likely to endanger the entire project.

I have focused so intently on the academic system because the report "The Case for Change" clearly recognizes the problem and because the compartmentalization into disciplines persists after graduation. Thus, in regulatory agencies, industry, research organizations, consulting firms, and the like, the disciplinary bias acquired in our educational institutions results in fragmented decisions, failure to integrate information that should be, and other unacceptable activities in the context of large-scale problem solving. The isolating mechanisms set up by the disciplines, such as different rituals for getting degrees, publication in speciality journals, development of a tribal language (an uncharitable person would say "jargon"), physical isolation in buildings, meetings, etc., and finally, insistence that all work be done in the particular way required by the discipline, simply are not acceptable where integrative studies are required. Until this problem in the educational system is corrected, it seems highly unlikely that it will be corrected in the other institutions upon which society depends (industry, state agencies, and so on). Even if it were possible, it would require expenditure of money and time undoing the work for which much money was expended in "educating" the students. If the educational system makes them less fit to work in a global environment, something is badly wrong with it.

FUTURE NEEDS

Despite the clear identification of the problem in "The Case for Change," strong resistance is likely to come from faculty who see a disciplinary orientation not as an artificial construct but as a way of life. Therefore, we must be certain that the rest of the world insists that all students have the opportunity to acquire the integrative capabilities required for decision making in this complex world. The students will quickly see where the career opportunities are and go to those institutions where integrative studies are not hampered by the tyranny of the disciplines. Over a relatively short period of time, the marketplace should adjust the system if integrative studies are required in businesses, governmental agencies, and the like. At the same time, councils of higher education should put pressure on universities to facilitate integrative studies.

Unfortunately, a common tactic on some campuses is to require that the interdisciplinary units make an attempt to structure their activities so that they are "congruent with the goals and aspirations of the departments or disciplines." Success in IEM will only come when the academic system becomes congruent with global problem-solving needs. *The problem should not be adjusted to fit the artificial compartments of the academic system, but rather the academic system must change to fit global problem solving needs!* Unfortunately, this simple fact has escaped most academic institutions.

A STUDENT'S POINT OF VIEW

I regularly get letters from students who wish to attend the graduate program at Virginia Tech for a degree in environmental studies. One that arrived recently is illustrative of a very general problem. The student has a B.S. in geology (shades of Charles Darwin), but was primarily interested in forest restoration ecology. Quite understandably, his Graduate Record Examination scores did not show the strengths in biology that would be expected had his degree been in that field. Nevertheless, the scores were quite respectable. He could get admitted to the Division of Forestry at Virginia Tech but would have to take some makeup courses in that area because his undergraduate degree was in another field, or, if he chose to go to biology, he would have to take makeup courses in that area. Geology, of course, where he would need no makeup courses, has no program in restoration ecology. The student's background is actually quite suitable for restoration because before an area is revegetated it must approximate its predisturbance condition in structural terms. The departments were unwilling to co-sponsor this degree (according to the student) because each department wanted its particular disciplinary specifications to be met. So, here is a student with interests in a topic of global concern (and with many theoretical aspects) with a background partly suitable to that concern who wants to add to it the particular talents necessary to make restoration of a terrestrial system possible. Our academic system instead tries to fit this student into the mold of a particular discipline, none of which is solely suited for the research problem he wants to undertake. The student will end up with a dissertation, if he chooses to get an advanced degree, that is congruent with the needs of the discipline but not with the needs of the research problem or the student's strongest interests. *Surely an academic system that is not more flexible is inappropriate for these times!*

EFFECTS OF GOVERNMENT REGULATIONS ON INTEGRATED ENVIRONMENTAL MANAGEMENT

Regulations, both federal and state, are often highly prescriptive. For example, in some states, there is an upper limit for temperature increase of water. While it is true that an increase to an elevated level would be disastrous for cold water fisheries, there are warm water fisheries where it is highly probable that no deleterious observable biological effects would be recorded with modest increases. In short, the scientific justification for applying a specific explicit temperature threshold for every type of ecosystem simply does not exist. There are ways to get around this, most notably Section 316 of Public Law 92500. Nevertheless, the persuasiveness of such evidence to regulators is uncertain and, therefore, this limit is often met when, in fact, there may be persuasive ecological evidence that shows it is arbitrary and produces no dramatic biological benefits for a particular ecosystem. The situation is worse than this, however, because people often even refuse to think along certain lines because government regulations prohibit use of certain kinds of evidence. If the regulations do not prohibit certain kinds of evidence, they often may not favor it or explicitly state how it will be used. In these doubtful areas, some people refuse even to consider alternative management courses of action that might make good sense in terms of theoretical ecology, but not in terms of existing regulations or perceived interpretation of these regulations. Thus, management options are strongly influenced. If IEM is to succeed, the well-being of the ecosystem must be the major determinant of all courses of action, not how the courses of action or thinking patterns fit arbitrary regulations not justified by scientific evidence.

Government bidding requirements have an interesting effect on the kinds of information gathered for IEM. For example, there is no robust body of scientific evidence showing that the same number of samples will produce the same results in different microhabitats or different ecosystems or even different regions of the same ecosystem. Nevertheless, I have encountered extraordinary resistance to the idea of taking samples until there is some evidence of information redundancy such as an asymptotic curve.[3] The people who resist this idea are well aware of the abundant literature justifying determining how much evidence is required to produce information redundancy, but state that requests for proposals (RFPs) and other bids for contracts sent to consulting firms and the like must specify a particular number of samples for each location (e.g., four) and that RFPs where the number of samples to be taken is not explicitly stated are intolerable. Thus, management decisions based on this evidence will be determined more by the need to fit the requirements of an RFP than the scientific requirement for good evidence. I have been told that it is not impossible to send out a bid where the exact number of samples is determined by the method just briefly described, but there is an extraordinary reluctance to do so nevertheless. Here is a prime illustration of a barrier to IEM; namely, that the evidence upon which the management decisions are made is influenced more by auditing requirements and bidding requirements than by obtaining reliable information about the condition of the system being managed.

CONCLUDING STATEMENT

Last summer, before the conference on IEM was held, I was privileged to read a draft of a manuscript by Craig Loehle on creativity in research. The most salient observations have been published in *BioScience*[4] and a citation on pp. 124 of that article indicates that a larger manuscript has been submitted and is in review. Loehle states that there are four requirements for a successful career in science: knowledge, technical skill, communication, and *originality or creativity* (italics mine). *He also makes the case that in some problem solving areas more creativity is needed than in others because phenomena are complex and multivariate.* Although he did not explicitly say so in the article, this is a perfect description of research problems requiring interdisciplinary information for successful resolution.

The following quote from Loehle[4] illustrates the problem of a disciplinary bias quite well:

> "A simple test for creativity involves giving test subjects a set of objects and a goal, to see if they can use ordinary objects in unusual ways (e.g., a rock as a hammer). Noncreative individuals are often stumped by such tests. In science, too, objects become fixed in meaning. In many cases, an assumption comes to have the rock hardness and permanence of a fact. My children had been playing with some yarn for months, calling it spaghetti for their toy kitchen. Whey my four-year-old daughter started twirling it around to the music, one piece in each hand and like the Olympic gymnasts, my five-year-old daughter became upset because you do not twirl spaghetti around and dance with it. Therefore, young scientists or those venturing in from other fields often make the most revolutionary breaks with tradition: they are able to ask, 'Is this really spaghetti?' "

In Figure 1 of Loehle's paper, he shows the relationship between the degree of difficulty and payoff from solving a problem, indicating that the payoff is severely diminished per unit of effort as the difficulty increases. This is what I have been trying to say throughout this chapter; namely, that integrative studies are made more difficult by obstacles originating in the disciplines. Creativity in solving a complex and multivariate problem that transcends a single discipline is therefore severely repressed, perhaps even extinguished, by the disciplinary biases cultivated in educational institutions and fostered in regulatory agencies and industry. The prejudice against interdisciplinary partnerships on most university campuses may be quite strong but will vary from department to department. It is interesting to speculate whether a candidate for promotion and tenure in Discipline X, who was also a spouse in an interracial marriage, would be more likely to be denied advancement because of the marriage or an interdisciplinary partnership. On most campuses, I suspect the latter would be more important in the denial. In neither case, of course, would there be overt recognition of the prejudice.

As is always the case, the penalties for any form of prejudice affect both the oppressors and the oppressed, and society suffers as a consequence. Academic institutions that pride themselves on an enlightened view racially are nevertheless often guilty, at least in some compartments of the institution, of flagrant prejudice against intertribal (read: *interdisciplinary*) collaborations. Fortunately, there is an increased recognition of the need to overcome the prejudice against interdisciplinary activities, which, regrettably, is often simultaneously tainted by being applied. Even on a land-grant university campus with a motto "That we may serve," I often find faculty members boasting that they are theoretical, not applied. It is no disgrace to solve a problem that will benefit humanity! If it is done creatively, the theoretical benefits should be abundant as well. I am frequently impressed by the novel approaches that nonecologists can provide to investigate "routine" ecological problems (or that geology student, untainted by a standard biological background, might provide to forest restoration). Interdisciplinary interaction helps to expose misguided preconceptions and assumptions of each investigator and, thus, increase the rate of problem solving.

ACKNOWLEDGMENTS

I am indebted to B. R. Niederlehner, David R. Orvos, Jay Comeaux and Robert B. Atkinson for reading and commenting on a draft copy of this manuscript. I am indebted to Teresa Moody for processing the initial dictation and several drafts on the wordprocessor, and to Darla Donald, Editorial Assistant, for preparing the manuscript for publication in this volume.

REFERENCES

1. Cairns, J., Jr. "The Emergence of Global Environmental Awareness," *J. Environ. Sci. (China)* 2(2): 1-7 (1990).
2. Maguire, L. A. "Decision Analysis: An Integrated Approach to Ecosystem Exploitation and Rehabilitation," in *Rehabilitating Damaged Ecosystems*, J. Cairns, Jr., Ed. (Boca Raton, FL: CRC Press, 1988), pp. 105-122.
3. Cairns, J., Jr., and J. R. Pratt. "Developing a Sampling Strategy," in *Rationale for Sampling and Interpretation of Ecological Data in the Assessment of Freshwater Ecosystems, STP894*, B. G. Isom, Ed. (Philadelphia: American Society for Testing and Materials, 1986), pp. 168-186.
4. Loehle, C. "A Guide to Increased Creativity in Research-Inspiration or Perspiration?" *BioScience* February:123-124 (1990).

CHAPTER **19**

Summary of Perspectives on Integrated Environmental Management

Laurence R. Jahn

This discussion provides some perspectives and conclusions gleaned from the presentations and discussions at this timely conference. Experiences in other watersheds and river basins are folded in to help provide insights for visualizing what is needed in the Savannah River Basin and at the U.S. Department of Energy's Savannah River Site.

The three Working Groups reached several common conclusions:

- Planning and management must focus on the entire Savannah River Basin as a functioning air/land/water system. In particular, hydrologic systems must be understood and human activities aligned to perpetuate qualities, quantities, flow characteristics, and essential relationships of surface waters, groundwaters, freshwaters, and brackish/salt waters. In the absence of appropriate intensive management, unbridled, unguided human activities will lead to degraded water conditions, losses of public values associated with the resource base, and costly restoration of degraded situations. Newspapers and other media outlets carry stories of far too many degraded units of the landscape, such as the Great Lakes, Chesapeake Bay, Florida Everglades, and others.
- Human activities in the Savannah River Basin are having impacts on the Savannah River National Wildlife Refuge and other areas in South Carolina. Those adverse impacts are signals that integrated environmental management is needed at multiple scales (e.g., entire basin, each watershed, site, etc.). For wide-ranging wildlife, such as the wood stork, migratory birds, and large predators, an area larger than the Savannah River Basin will have to be used in practical planning and management.
- To handle integrated environmental management on multiple scales will require new, stronger institutional arrangements to (1) perpetuate resources and values covered

under the public trust doctrine of law and (2) plan, manage, and monitor management efforts and other human activities, frequently involving mixed, rather than single, land ownerships. Management is required to prevent activities incompatible with ecological processes, avoid associated restoration costs, and maintain the environment in productive condition to support long-range sustainable uses. Simulation models are useful to plan and monitor waterflows and other characteristics of the Basin's river systems at multiple scales to help ensure the integrity of ecological processes and communities. However, are natural resource legal authorities, policies and regulations, as well as employee job descriptions and performance criteria and standards, aligned to achieve appropriate compatible developments? Likely not in many cases. Constructive realignment of such laws, guidelines, and procedures is essential if accelerated progress is to be achieved in implementing ecologically sound integrated environmental management.

- Within a framework of facts on natural resources and the ecological processes to be perpetuated, people should be given choices (alternatives) to carry out their activities compatibly with sustained functioning of the resource base. An improved base of facts is needed for the Savannah River Basin and Site to permit clear identification of alternatives for future compatible uses. Those plans should be oriented to provide people with pleasant, safe places to live, as well as socioeconomic conditions that provide opportunities for achieving and maintaining a reasonable standard of living.
- While ethnic groups of people have a variety of traditions and views on using the resource base, their needs and desires should be satisfied with and through ecologically sound plans and practices. That framework of ecological soundness is crucial.

Generating these plans will require interdisciplinary teams that deal with a variety of scales, some larger than the Savannah River Basin, as well as other site-specific ones. This multiscaled perspective will require planners, engineers, managers, and others to be far- as well as nearsighted in their perspectives.

Some examples illustrate the types of approaches needed in the Savannah River Basin, or any other river basin or watershed, to perpetuate ecological functioning systems:

- Limit or eliminate inflow of chemicals (e.g., nitrates, phosphates, pesticides, radioactive materials, etc.) to surface waters and groundwaters.
- Guide removals of water from aquifers to maintain balance with recharge. For example, limit time and volume of pumping to prevent adverse impacts on surface waters (e.g., streams, lakes, wetlands, etc.) and avoid land subsidence and salt water intrusion in coastal areas.
- Delineate and protect groundwater aquifer recharge areas. Laws in Iowa and Wisconsin are among the best available on groundwater. They could be helpful in stimulating new approaches to improve management of aquifers in the Savannah River Basin.

For forests in the Basin, it is absolutely essential to have inventory data for the types and distributions of vegetative communities. With the current status of each forest type established, it is then necessary to define the desired future condition for each type and design or prescribe practices to achieve that desired future status. The challenge is to design and use tree harvest/regeneration prescriptions to ensure sustained yields of multiple values and benefits, not solely wood yields. Accompanying information must identify which wild living resources are advantaged and which are disadvantaged and for how long under the various prescriptions.

For agricultural lands, inventory findings should show the soil types and acreages of various commodities produced in the Basin, as well as current rates of soil erosion and use of chemicals, such as fertilizers, pesticides, etc. Rates of soil loss should not exceed acceptable or tolerable limits, such as 3 or fewer tons per acre per year on shallow soils and 5 or fewer tons per acre per year on deep soils. Those limits should serve as guides for correcting excessive erosion situations and for preventing accelerated noninduced erosion. With the United States having to take 100 to 150 million acres out of commodity production, there is no reason now to encourage converting wetlands, forests, or grazing lands to croplands.

For rivers, lakes, reservoirs, and wetlands, management of river flows must be orchestrated among the reservoirs to perpetuate the ecological processes, minimize and avoid adverse impacts of salt water intrusion, and ensure delivery of adequate quantities and qualities of water to fish/wildlife/recreation management areas. Floodplains and shorelands of 100 to 1000 ft from the high or average water levels should be zoned and managed intensively. Only compatible uses should be permitted. This integrated management is critical to avoid further increases in mounting taxpayer-funded disaster relief payments.

For unique areas in the Savannah River Basin, special management treatments are required. Examples include:

- Groundwater recharge areas. They need to be delineated and prescriptions should allow only human activities compatible with their vital functions.
- Contaminated aquatic and terrestrial sites. Contaminants, together with their locations, must be mapped to improve understanding of site employees and the general public of risks involved and potential future uses of the sites.
- Research areas. Existing weak data bases emphasize that areas are required where studies can be pursued for some years without being confounded by a variety of incompatible human influences.
- Habitats of sensitive, threatened, and endangered species. Actions should be taken to improve fish and wildlife habitats and populations. Goals should be to keep sensitive species from becoming threatened or endangered, and encouraging recovery and associated delisting of threatened and endangered species.
- Wood stork foraging areas, such as the delta near one creek. These unique areas must be given priority attention.

Realignment of perspectives and actions to improve management of natural resources was interwoven with three common threads:

1. Economic evaluation procedures and results can lead to inappropriate decisions and actions. Calculation procedures encompass dollar values, but set aside important ecological and other values beyond dollar expression — labeling them "externalities."

Although "bagged" and set aside, the values remain, but frequently are ignored. Among those values are sustained functioning of ecological systems, quality of life, and options for the future. Without full consideration of these values in decision making, bad decisions can be and have been made.

Fortunately, a few economists, such as Herman Daly of Louisiana State University, now are rethinking the calculation procedures and adding ethical and ecological criteria. This is an important advance to bring the relationship between the economy and environment into clearer focus and to place decisions and activities on a sustainable, rather than an exploitable, basis. Collective actions will be required within communities, watersheds, and river basins. Without such cooperative, coordinated actions, there is a high probability of environmental damage resulting from individual decisions and actions that remain unchecked by community guidelines.

Growth must be compatible with characteristics of landscape ecosystems, such as those of the Savannah River Basin. Integrated environmental management seeks to establish sensitive uses of the resource base on a sustainable, multibenefit basis. Only through application of environmental and social criteria and standards can that goal be achieved. Doing what is right, rather than outright exploitation, must prevail. Some policies and procedures are in place; others remain to be crafted and implemented.

2. Ethics are important to help guide actions. While several perspectives on ethics were discussed at the conference, a few additional aspects need attention to strengthen the framework within which to consider the Savannah River Basin or any other unit of the landscape.

When Aldo Leopold's classic essay on the land ethic was published in 1949, it was obvious that he saw, as few others did at the time, that the life support systems of the landscape were breaking down. The interdependent relationships of air, soil, water plant, animal, and people were stressed and degrading. Leopold knew the scientific reasons for it, but suggested a more fundamental cause: our way of thinking about the land (resource base) and our relationship to it was deeply flawed. If the health and productivity of the land, on which all living things depend, are to be sustained, ecological functional processes must be maintained. Common sense of doing what is right must prevail.

This message surfaced early in 1989 from the Bruntland Commission of the United Nations World Commission on Environment and Development. It called for a new global ethic to resolve conflicts between human activities and environmental health. The Commission stated: "We must learn to accept the fact that environmental considerations are part of a process for unified management of our planet. This is an ethical challenge."

It now seems clear. Society must be circumscribed by the limits of ecosystem health (Earth's processes and wild living resource health). Any individual, agency, or corporation that ignores this imperative will eventually become subjected to public scorn, leading to distrust.

3. The general public is concerned about the status of the resource base and wants to see improved management applied. Findings from various polls, combined with accumulated experiences, show that citizens are aware of resource problems and management needs. They are willing to support constructive actions. In California, a strong coastal zone management act was enacted through a ballot approved by the people. In Missouri, one eighth of 1% sales tax, as well as an additional one tenth of 1% sales tax, were established by majority vote of the people.

While these successes in California, Missouri, and other states are impressive, attitudes and strength of support for improving management of natural resources varies among states and counties or parishes or boroughs. Nevertheless, in many geographic areas, people, as well as elected representatives, are seeking constructive change. For example, in the Food Security Act of 1985, a conservation dimension was woven or integrated into agricultural commodity programs. Parallel actions, with emphasis on water, are anticipated in the 1990 Farm Bill. These cases show what can be accomplished when innovative resource management plans are supported firmly by the public.

It is most fitting that we assembled here in Aiken to explore and respond to scientific, ethical, and practical resource management needs focused specifically on the Savannah River Basin and Site. This is precisely the type of attention needed throughout the United States and elsewhere. Hopefully, publication of this group's collective thoughts will stimulate needed attention and constructive actions in other river basins and watersheds. Enlightenment and responsive actions will depend on well-organized efforts to work with the public and decision makers.

A specific follow-up plan will be required to ensure that proposals and recommendations from this conference are implemented to improve integrated management of the Savannah River Basin and Site. That game plan awaits development.

REFERENCE

Leopold, A. *A Sand County Almanac* (Cambridge, MA: Oxford University Press, 1949).

APPENDIX I

Principal and Illustrative Questions for the "Management of the Savannah River" Working Group

How does one develop and implement an integrated environmental management plan for the Savannah River that meets most of the needs associated with the uses of the river?

ILLUSTRATIVE QUESTIONS:

What is the chemical and biological "health" of the Savannah River?

How does the Corps of Engineers decide the acceptable level of the Savannah River during times of drought?

How are the needs balanced for water of the property owners on the shores of the up river reservoirs for recreation, tourists, and aesthetics with the down river water users (the SRS, public drinking water plants, river boat traffic, etc.) and with hydroelectric power generation?

What group should receive priority — up river reservoirs or down river users? What groups should be involved in setting criteria? What factors are involved in criteria setting and how are they weighted?

What is the regulatory action against discharges to the river when low river flow causes them to exceed their NPDES permit mixing zone requirements?

What consideration should be given to wetlands as a user of down river water? In a priority sense should river and estuarine wetlands be weighted more or less than reservoir shoreline wetlands?

What role should the fisheries of the river and of the reservoirs play in water level considerations?

Who should evaluate the total impact of all effluents to the Savannah River below the dams from South Carolina and Georgia? Who should then control the permitting action for the health of the river?

Who resolves disputes among different states, different federal agencies, local governments, industry, and "environmental" groups?

Should the river stage be maintained at a near constant height below the dams or should it be allowed to fluctuate as determined by usage demands? What is best for the wetlands and how much is it worth?

Are there endangered species considerations in the management of the Savannah River? If so, what value should they be given in the management strategy?

How many more demands can be placed on the uses of the Savannah River — as a source of water and as a receptor for industrial and domestic waste?

What impact will the City of Savannah increasing its use of the Savannah River have on the management of the river?

Is salt water intrusion up the river a problem in the Savannah River estuary? If so, how is it balanced with the salt water intrusion in the groundwater in the City of Savannah area and the city's plans for using more of the Savannah River for a drinking water source?

Are there other factors besides those indicated above that need to be considered in making environmental management decisions for the river?

What process should be in place for making environmental management decisions for the Savannah River?

Principal and Illustrative Questions for the "Endangered Species Protection — The Wood Stork Example" Working Group

What is the magnitude and type of effort that should be expended in protecting endangered species?

ILLUSTRATIVE QUESTIONS:

What is the rationale for protecting endangered species?

How are the benefits of endangered species protection balanced with the costs of doing so?

Should endangered species be introduced into new areas? Are we obligated to protect/nurture them if conditions are not optimum for their survival?

Should the water level of the Savannah River be controlled in such a way during the late spring/early summer feeding season to provide food for the wood stork?

What was the historic role of the "ox bows" and flood plains in providing feeding areas?

How much is it worth per bird to protect the wood stork by providing feeding areas?

Is it reasonable to expect the provision of feeding areas to ensure the survival of the local colony?

How long should alternate feeding be provided by managing Kathwood Lake for that purpose?

Is the survival of the Birdsville, GA, colony important? Should the predators be controlled?

What are the relative roles of the nesting, roosting, travel routes and feeding areas on the survival of the wood stork?

If survival of the local colony is important, who should be responsible for all aspects of environmentally managing it? How should it be done?

Principal and Illustrative Questions for the "Long-Term Management of Savannah River Site Lands" Working Group

What should be the strategy for the SRS site after the nuclear materials production facilities cease operation?

ILLUSTRATIVE QUESTIONS:

What should be the long-term use of the lands — environmental research park, national forest, returned to the public, permanent waste disposal site, industrial development site, etc., or some combinations?

What should be done with the nonoperational facilities?

Should some facilities/locations be protected forever?

Is it environmentally and cost effective to meet all aspects of state and federal environmental laws and regulations for parts of the SRS site that would be protected forever? What strategy should be followed if not environmentally and cost effective?

Is it advantageous to dismantle and clean up some of the major facilities and waste sites? If not, how should they be preserved and protected?

What strategy should the SRS be following now to be consistent with the long-term strategy developed?

APPENDIX II

Speakers and Participants

Dr. Bud Badr
1201 Main Street
Suite1100
South Carolina Water Resource Commission
Columbia, SC 29201
Phone: 803-737-0800

Mrs. Susan Bloomfield
500 Norwich Road 40
Augusta, GA 30909
Phone: 404-736-4348

Dr. I. Lehr Brisbane
Savannah River Laboratory
Drawer E
Aiken, SC 29802
Phone: 803-725-2472

Dr. John Cairns, Jr.
University Distinguished Professor
Department of Biology
Director, University Center for Environmental & Hazardous Materials Studies
1020 Derring Hall
Virginia Polytechnic Institute and State University
Blacksburg, VA 24061-0415
Phone: 703-231-7075

Mr. Fred B. Christenson
3 Ivy Circle
Aiken, SC 29801
Phone: 803-648-4998

Mr. Jerry J. Cohen
Science Application International Corp.
3015 Hopyard Road
Suite M
Pleasanton, CA 94566
Phone: 415-463-8111, ext. 107

Dr. Malcolm C. Coulter
Savannah River Ecology Laboratory
Drawer E
Aiken, SC 29802
Phone: 803-725-2472

Dr. Todd V. Crawford
Westinghouse Savannah River Company
Savannah River Laboratory
Aiken, SC 29802
Phone: 803-725-2767

Dr. Peter M. Cumbie
Duke Power Company
P. O. Box 33189
Charlotte, NC 28242
Phone: 704-373-7867

Dr. Kenneth L. Dickson
Director, Institute of Applied Sciences
University of North Texas
P. O. Box 13078
Denton, TX 76203
Phone: 817-565-2694

Mr. Jack C. Dozier
Chief, Water Protection Branch
Georgia Department of Natural Resource
205 Butler Street, SE
Atlanta, GA 30334
Phone: 404-656-4708

Dr. Dan E. Dykhuizen
Department of Ecology & Evolution
Division of Biological Sciences
State University of New York at Stony Brook
Stony Brook, NY 11794-5245
Phone: 516-632-8600

Dr. Peter Frederick
Wildlife & Forestry
118 Newins/Ziegler
University of Florida
Gainesville, FL 32661
Phone: 904-392-4851

Dr. Elizabeth C. Goodson
Department of Energy
Savannah River Operations Office
Aiken, SC 29802
Phone: 803-725-3965

Dr. Judy Gordon
Department of Biology
Augusta College
2500 Walton Way
Augusta, GA 30910
Phone: 404-737-1539

Dr. Larry Harris
Wildlife & Forestry
118 Newins/Ziegler
University of Florida
Gainesville, FL 32661
Phone: 904-392-4851

Mr. John G. Irwin
U. S. Forest Service
Savannah River Forest Station
Building 760-G
Aiken, SC 29802
Phone: 803-725-2441

Dr. Lawrence R. Jahn
President, Wildlife Management Institute
1101 14th St., NW
Suite 725
Washington, DC 20005
Phone: 202-371-1808

Mr. J. Richard Jansen
U.S. Department of Energy
Savannah River Operations Office
Aiken, SC 29802
Phone: 803-725-1425

Dr. Bernd Kahn
Environmental Research Center
Office of Interdisciplinary Programs
Georgia Institute of Technology
Atlanta, GA 30332
Phone: 404-894-3776

Dr. Randall A. Kramer
School of Forestry & Environmental Studies
Duke University
Durham, NC 27706
Phone: 919-684-6090

Mr. Ronald A. Lanier
Assistant Chief, Planning Division, Savannah District
U. S. Army COE
U. S. Army Engineer District
P. O. Box 889
Savannah, GA 31402-0889
Attention: CE SAS PD-X Phone: 912-944-5808

Ms. Susan C. Loeb
South Eastern Forest Experiment Station
Department of Forestry
Clemson University
Clemson, SC 29634
Phone: 803-656-4865

Professor Margaret N. Maxey
Biomedical Engineering
Chair of Free Enterprise
Petroleum/CPE 3.168
University of Texas at Austin
Austin, TX 78712
Phone: 512-471-7501

Dr. William D. McCort
Savannah River Ecology Laboratory
Drawer E
Aiken, SC 29802
Phone: 803-725-2472

Dr. Ian McHarg
Chairman, Department of Landscape Architecture
  and Regional Planning
University of Pennsylvania
34th and Walnut Streets
Philadelphia, PA 19104
Phone: 215-898-6591

Dr. Halkard E. Mackey
Savannah River Laboratory
Westinghouse Savannah River Company
P.O. Box 616
Aiken, SC 29802
Phone: 803-725-5924

Ms. Nora A. Murdock
U. S. Department of the Interior
Fish & Wildlife Service
Endangered Species Field Station
100 Otis Street, Room 224
Asheville, NC 28801
Phone: 704-259-0321

Dr. Thomas M. Murphy
Nongame Endangered Species Section
South Carolina Wildlife & Marine Resource Department
Rt. 2, Box 157
Poco Sabo Plantation
Green Pond, SC 29446
Phone: 803-844-2473

Dr. Eugene P. Odum
Institute of Ecology
University of Georgia
Athens, GA 30602
Phone: 404-542-2968

Dr. John C. Ogden
Acting Program Manager, Wildlife
U. S. Department of the Interior
National Park Service
Everglades National Park
South Florida Research Center
P.O. Box 279
Homestead, FL 33030
Phone: 305-245-1381

Dr. Robert V. O'Neill
Environmental Science Division
Building 1505
Oak Ridge National Laboratory
Oak Ridge, TN 37831-6038
Phone: 615-574-7846

Mr. Morrison J. Parrott
Executive Director, Savannah Valley Authority of South Carolina
P. O. Drawer K
McCormick, SC 29835
Phone: 803-391-2410

Dr. Ruth Patrick
Francis Boyer Chair of Limnology
The Academy of Natural Sciences
19th and The Parkway
Philadelphia, PA 19103
Phone: 215-299-1098

Mr. John F. Proctor
Consulting Services
Proctor & Associates
Specializing in Nuclear Energy
106 Englewood Drive
Aiken, SC 29801
Phone: 803-648-6866

Dr. Lee E. Rodgers
Biological Sciences Department
PNL
Richland, WA 99352
Phone: 509-376-8256

Mr. Daniel P. Sheer
Water Resources Management, Inc.
Hillcroft 1, Suite 210
6310 Stevens Forest Road
Columbia, MD 21045
Phone: 301-381-1881

Mr. Peter Stengle
Savannah River Ecology Laboratory
Drawer E
Aiken, SC 29802
Phone: 803-725-2472

Dr. Robin L. Vannote
526 West Street Road
Kenneth Square, PA 19348
Phone: 215-869-2784

Ms. Nancy Weatherup
Environmental Engineer, Bureau of Water Pollution Control
South Carolina Department of Health & Environmental Control
2600 Bull Street
Columbia, SC 29201
Phone: 803-734-5246

Dr. F. Ward Whicker
Department of Radiology & Radiation Biology
Colorado State University
Fort Collins, CO 80523
Phone: 303-491-5343

Mr. Steve R. Wright
U. S. Department of Energy
Savannah River Operations Office
Aiken, SC 29802
Phone: 803-725-3957

Dr. Paul B. Zielinski
Water Resources Research Center
Clemson University
310 Lowry Hall
Clemson, SC 29634-2900
Phone: 803-656-3273

Index

acid rain 65, 93
alkaloids, role in biodiversity 176
applied ecology 63—70
   ambiguity of ecotoxicology 68—70
   approaches in 65—66
   difficulty of 70
   importance of disasters in 66—68
applied research 96, 180

Bhopal 30
biodiversity 158, 180
   alkaloids in 176
   erosion of 103, 107—108, 111
     "fauna" vs. "wildlife" 103—104, 112
   "hidden" 82
   hierarchal approach in 112
   multiple use modules as strategy in 113
   preservation of 177, 178
   reductionism in 101
   as risk to survival of civilization 119
   role of ESA in 99—113
biogeochemical system 44
biohazards 28
biological succession 91
biomass 45

carcinogens, threat to human health 27—29, 74, 75
Carson, Rachel 25—27

"balance of nature", view of 26, 29
Chernobyl 30
clean-up of contaminated lands 71—77, 90—92, 180
   biological succession approach in 91
   Exxon Valdez oil spill 64, 67, 71
   hazardous waste internment 92
   historical assessment 74—75
   objectives of 72—73
   radionuclide levels in 90—92
   zero contamination 88
clearcutting 46, 81, 120, see also rain forests

DA, see decision analysis
decision analysis (DA) 9, 53—60, 159, 185
   groundwater quality management 58—59
   models
     expected utility 54, 57—58
     hierarchal goal structure of 58
     safety first 54, 57—59
decision tree 55—57
   management of endangered species as 56
   rehabilitation of damaged ecosystem as 56
defensive environmental management 8
disasters 8, 12, 29—30

confrontations with 66—68
need for controls in 67

ecological succession 82
economic risks 56
ecosystems
  energy flow function in 46
  health of 63—65; see also applied ecology
  lotic 45
  reciprocal design concept of 81
  stability of 122, 178
  use of radionuclides to chart movement in 83
ecotoxicology
  ambiguity of 68—70
  human health as indicator of 64
  surrogates in 69—70
  use of SRS as study site 95
Endangered Species Act (ESA), role in biodiversity 99—113, 122—124, 177
  as multiple use module 113
  population management 102—103
  preservation ethic 101, 112
endangered species, protection of 121, 157, 170, 175—178
  ethics of 176—177
energy-processing system 44
environmental risks 18, 22, 25—32, 58, 159, 163
  assessment of 33, 34, 59, 68, 74, 88
  environment as culprit 28—31
  environment as victim 26—28
  perceived vs. actual 32
  subjective probability 54
  zero probability 54
ethical plumb lines, managing risk using 21—36
  macro-ethics 35—36
  micro-ethics 35
  molar-ethics 35
  normative standards in 23
ethics 22—25, 196

of endangered species preservation 176—177
reductionism in 33
eutrophication 147, 148
Everglades National Park 109
  competition between species for dominance in 110—111
  endangered species living in 104
  extinct species 111
  mortality rates 105
  native species, extinction threat from non-native species 107
  nutrient cycling in 106
  proposals for future management 193
  use of systems approach in 108
Exxon Valdez 64, 67, 71

faunal collapse, southeastern United States 103—108, 177; see also Everglades National Park
faunal relaxation, see faunal collapse
fire management 42, 43
fire prevention 42, 73
flood storage volume 153

global warming 7, 8, 15, 82, 93, 185
  atmospheric carbon dioxide in 7
government regulations, effects of on integrated environmental management 189
groundwater
  aquifers 147, 154
  contamination of 89
  flow patterns 92
  management 58—59
  recharge areas, future management treatments for 195
  use of dams to recharge 148, 168, 171

hazardous waste 76, 96
  internment of 92
  limiting inflow of 194
  risk in disposal of 74

INDEX   213

saccharine as 76
use of SRS as storage site for 95
HERMES model as inappropriate assessment scale 42
holistic view of integrated environmental management 13, 43—46, 185; see also systems analysis
hummingbirds, spread of radiation by 84
hydroelectric power 147, 152, 154, 160, 164, 171
  pumped storage 154
  reduced 170

indigenous biota, impact of extensive irrigation on 7
Industrial Revolution, impact of 27

Love Canal 29, 73

man-made pollutants 29, 34
mesocosms 84
migratory birds 84, 193
Mycteria americana, see wood stork

National Environmental Research Park (NERP) 80, 89, 93
natural resource system, surveillance of 8
NERP, see National Environmental Research Park
nondegrading assimilative capacity of environment 10
nutrient cycling, use in ecosystems 44, 45, 79, 106

old growth forests 81, 100, 108, 112

PCB contamination 152
pesticides 53, 194
plant and animal pests, damage to environment 6
plutonium 83—84
policy decisions 21—23, 33, 59, 65

macro-ethics in 35—36
Prince William Sound, impact of oil spill on 8

radioecology 180
radionuclides
  levels in contaminated lands 90—92
  use in ecosystems 83
rain forests 6, 7, 15, 120
  deforestation of 122, 158
  effect of acid rain on 65, 93
reciprocal design concept 81
reductionism 13—14, 33, 101, 158, 185
restoration technology 95
risk and uncertainty, importance in integrated environmental management 54, 59—60

SA, see systems approach
Savannah River, see also Savannah River Site (SRS)
  dams on 147, 151
  eutrophication in 147
  flooding of 152—153
  flora and fauna of 141—147, 170
  history of 137—139
  management of 161—172
  fish spawning in 168
  scaling, in models 46
  confusion of 43
  incompatibility of 43
  perturbation 43
  Smokey the Bear 41
  spatial 43
  temporal 43
Savannah River Site (SRS) 79—84
  as microcosm of global ecosystem 178
  as wood stork feeding ground 127—128
  biodiversity in 101
  dams at 151, 162
  demand on groundwater aquifers 154

designation as NERP 80, 89, 93, 180
drought in 151, 153, 163, 171
　effect on fish spawning 155
federal laws applicable to 2
groundwater contamination in 89
hazardous waste internment 92
impact of radioactive materials on 89
long-term management of 87—96, 179—181
management objectives 164, 170
options for future management 92—96, 193
PCB contamination in 152
presence of near-surface groundwater in 95
radiation effects in 83—84
radionuclide levels in 90—92
sediment-water interface 45
spiralling length, function in nutrient cycling 45—46
SRS see Savannah River SiteSuperfund 67—68, 76
surrogates
　microcosms as 69
　most sensitive species 68
　organisms as 69
　toxicological 69
systems approach (SA) in ecology 39—46, 159
　as holistic mindset 39, 43
　biomass in 45
　energy flow in 44
　　power 44
　methodology of 39—43
　models
　　assessment 43
　　ecological 40
　　HERMES 42
　　risk analysis in evaluating 40
　　scaling of, see scaling, in models
　　succession 40
　nutrient cycling 45—46

technological hazards 29—32
thermal perturbation 82
Three Mile Island 30
Toxic Substances Control Act 27, 28
"turf battles" between agencies 9, 14

wetlands 79, 83, 94, 126, 127, 147, 153, 154, 164, 194
wood stork (*Myceteria americana*) 119—133
　as endangered species 89
　causes for decline 125—126
　Birdsville, Georgia, colony, significance of 126—127
　Kathwood Lake, as replacement foraging habitat for 130—133
　natural diversity of, economic importance of 175
　potential population crash 110—111
　proposal for future management of 193
　Savannah River delta as feeding ground for 82, 127
　wetlands as foraging grounds for 127 195

"zero pollution", quest for 28